RHED

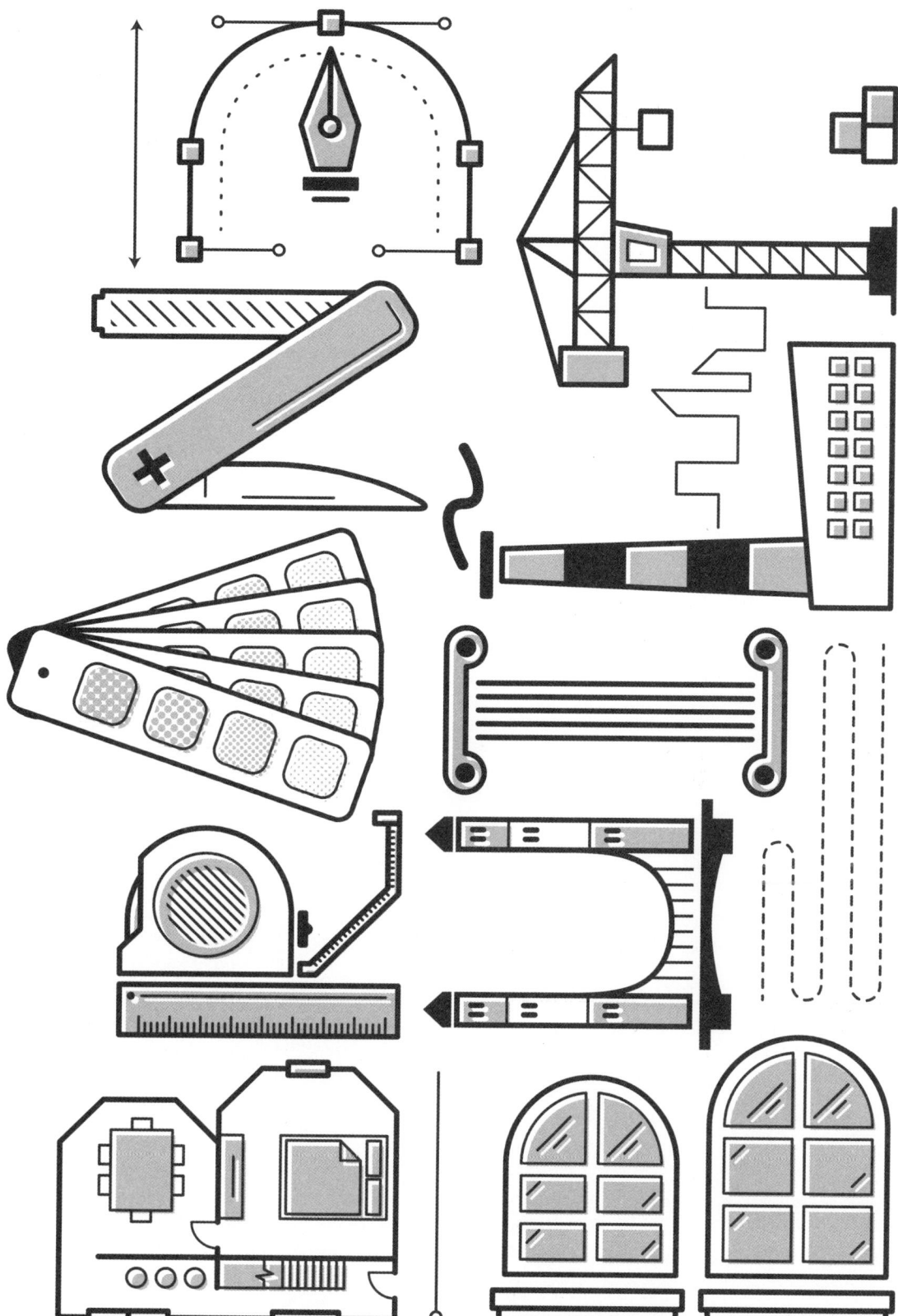

BRAND BUILT
Branding Buildings and the Builders

Tan Yang International Pte. Ltd.
50 Playfair Road, Noel Building, #07-02, Singapore 367995
Tel: +65 6289 9208 | Fax: +65 6289 9108 | Email: enquiry@tanyangintl.com

ISBN 978-981-09-9030-5
Printed by Tiger Printing (Hong Kong) Co. Ltd

EDITORIAL NOTES

- ≈ Projects are current as of March 2016.
- ≈ Please be informed that buildings may or may not be at the current stated location at time of printing.
- ≈ Some credits are intentionally omitted at the contributor's request.
- ≈ Company and brand names that appear in this book are copyrights and/or registered trademarks of their respective companies.

CONTENTS

1.2 Brand Identity

The brand identity is the overall distinctive look of the company or person, and extends beyond logo design to include its application on stationery, websites, advertisements and so on. When a brand identity works, one should be able to tell the brand even if the logo is not shown. For example, a sizzling red packet of fries will instantly remind you of McDonald's famous fries, even if the logo and name of the company is not shown.

While a logo cannot be changed, the brand identity should be consistent and flexible. The elements that make up the brand identity varies, and includes typefaces, choice of colours, choice of photography and imagery, as well as the copy-writing style used.

The stationery set (part of the entire brand identity) for Grupo BHAU.

Designed by Nhomada for Grupo BHAU (See page 124).

1.3 Brand

If the logo is the face of the company, and the brand identity is much like the choice of clothes that company choose to wear, then the brand is the personality and character that the company is associated with. In simple terms, the brand is how you want the company to be referred to as a person.

The brand is the overall process that encompasses the logo and the brand identity, and also includes other areas such as business positioning, the storytelling of the company's beginnings or operations, all the way down to the front-line customer service and the company's feedback channels. It is essentially the sum of all of the impressions, customer and employees experiences and the understanding of the public perceptions towards the company. Often, the company has no control over how their brands are being perceived, and is usually generated by how the public views and define the brand and the company. For instance, even with a fantastic logo and an excellent product, a bad review of the company by external sources could harm the brand, putting negative perceptions and impressions in the minds of the public.

The common mistake is that most people refer the logo as the brand, where in actual fact, the logo is part of the outcome of a more extensive process called branding. Aside from the logo and the brand identity, the branding process considers the company values and vision, and review how customers expectations and impressions match with what the company thinks of itself internally. Most companies view logo design as an aesthetic exercise with no regard for the company's vision, mission and overall impressions and perceptions, and find themselves having to re-brand and reposition themselves repeatedly over the years. Logo design and branding should work side by side — where behind every great logo that is creatively designed, stands a brand that is clearly defined and unique.

Differences between a logo, a brand identity and the entire branding.

Logo   Brand Identity   Branding

personality   attitude
behaviours   perceptions
charisma   feedback

Logo
Brand Identity
Branding

2 BRAND PROCESS

Essentially, the brand is like a person, and the logo is the face of the company. Without these, a company cannot operate — much like speaking to someone without a voice and without a face.

Given the importance of branding, where do we start this mysterious branding process? A branding process is one that leverage the potential of a great logo and matches it with the company's goals, mission, vision, values and perceptions. A start-up should consider this process carefully to define the focus and identify how you want the company to be portrayed.

For companies seeking a re-brand, it is a good time to review whether the brand identity of the company matches with your business operations, customer feedback channels and your tone of voice when communicating internally and externally. A re-brand is also conducted when the objectives of the company shifts, or when the logo is out of date. Sometimes, a re-brand could also occur as result of larger motivations and issues, such as acquisitions, mergers or the emergence of new market segments.

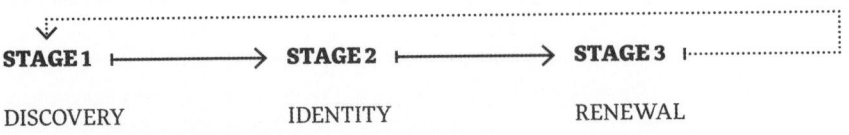

STAGE 1 ⊢————→ **STAGE 2** ⊢————→ **STAGE 3** ⊢·············

DISCOVERY   IDENTITY   RENEWAL

2.1 Discovery

Before any of the branding and fancy logo design process begins, we would have to first identify and review what has been done before and what is being done presently.

2.1.1 The Past

The past of every person shapes how the person thinks, perceive or behave. Similarly, brands also have historical records and stories of how these brands came to be.

Reviewing the company's history can help understand who you are, and better review the company's position even if you assume to know it. More often than not, what the owners and employees think about their companies differs greatly from how the public perceives the company. Assuming how the customer sees and perceives the company is almost like entering a war without a proper strategy. Even in war, we have to first understand our enemies, the terrain and many other factors before coming out with a proper and cohesive strategy.

For instance, a noodle shop that has consistently long queues is believed to be delicious by the staff of the shop. However, in actual fact, this is due to the price being the cheapest in the vicinity, and hence the queue. Branding this noodle shop as being delicious will convey the wrong idea to potential customers. If customers has the expectations that the noodles will be delicious, customers will be disappointed and negative impressions will be formed.

Besides the business and operational side of things, check and review past logos that the company have used, and understand what works and what doesn't, what should be retained and what should be removed. Also check against whether old logos matches the company's vision and mission.

2.1.2 The Present

Review the company vision and mission by defining what the company stands for, much like how you will set up a stall and differentiate it from the other vendors. This is also a good exercise to benchmark your current state of affairs against what the vision and mission states. Look at how the company is run and how it can be run differently from other similar stores, at the same time, review what your competitors have got it right.

An example will be a eco-friendly and fair trade tea distributor, with "ethical" as a value that the brand want to be associated with. This becomes a point of reference for all operational processes, from the procurement of fair trade tea leaves, to the decisions involving suppliers and distribution channels, to how the tea bags are being packaged using forest-friendly papers, and even to the extent of you deciding to employ unpaid interns.

One way we can review the brand is to think of the unique attributes that the brand wishes to portray itself — the unique selling point. A unique selling point might be about the products, or it could be a personality or characteristics that sets you apart from the competition. For instance, every spectacle shop sells spectacles, but companies like Owndays have repositioned and presented themselves for their unique way of producing the spectacles — fast, affordable, yet durable. A project manager, for instance, might be "detail-oriented", "reliable" and "a good communicator", but project managers can also be "humorous" to allow staff to be relaxed amid the pressure, and to get the project completed without any unhappiness.

Ideally, companies should be able to define the company in a single sentence. It is tough, but one of the ways to achieve that is to tell a story. Take for instance, the three sentences below:

▶   Fair trade coffee for all occasions
▶   Artisan coffee, specially made for you
▶   Artisan coffee made with love, using the finest hand-picked coffee bean from fair trade sources

From the above sentences, the last sentence tells a story although it is similar to the other two. "Artisan coffee," feels a little more class and emotional, as it implies that the coffee is artfully created for the consumer. The coffee is brewed by someone who is a barista and is someone who knows their craft well; the "finest hand-picked" suggests the best beans and adds to this special exclusive one-off feeling. "Made with love" suggests that the coffee is roasted and brewed meticulously and with delicate care.

Besides the internal review of the company's operations, vision and mission, companies should also identify their target audience. These target audience profiles should also be as detailed as possible: are your clients end user consumers or trade buyers, are they of a certain age group or gender, are they dog lovers, coffee drinkers, foodies, and so on. At this point, it is also an appropriate time to ask if you could do anything differently to increase your customer base. At the same time, think about how to retain your customers, find out how existing customers view your brand, and identify what you want your customers to view your brand and your company.

Finally when in doubt with the review process, get a trusted friend to see it. Sometimes, the most obvious mistake or assumption is easily identified through a third party point of view, and this could bring about some clarity and focus to your research.

2.2 Identity

After reviewing your company's internal perceptions and customer's impression, we move on to the next phase of the branding process — the visual identity. Armed with your research in the Discovery stage, the information will guide you into making rational decisions and not just by gut feelings. For instance, if your target audience are up-market people and you want to stand out as a artsy brand, a corporate yet rigid identity might not make the cut even though the logo and identity is well-designed.

2.2.1 Name

The toughest decision of every company — the name. To be honest, it is even harder to name your first child! There isn't any rule about how companies should be named, but it is crucial that we should not get too engrossed with the first name that was developed. Naming takes time and as it develops further, a more well-thought and appropriate name should surface.

Besides coming out with the name that you like and a name that matches the research done in the first part, it also important to find out how this name can be translated into a URL or a website address. See the example below, where we are creating a website address for this book:

► brandbuilt.com (not available)
► brandbuilt.org (available)
► brandbuiltbook.com (available but longer)
► bbbook.com (available but even shorter)

From the above, it seems that "brandbuilt.com" is the ideal name but it isn't available. The domain ".org" might not as easy to remember than a ".com" domain, and "brandbuiltbook.com" seems longer. Abbreviating it might lose its meaning and does not explains what this website leads to. However, if you abbreviate, it might be generated into a idea for a logo, and "brandbuiltbook.com" does seem easy enough to remember.

Besides checking your name online, check with your local government company registrar to see if there is another company that has the same name as you. There are many companies that are not registered and it is better to check your name against the registrar before committing to any name.

2.2.2 Logo

The logo is like a signature that embodies what the company and the brand believes in. It is a graphical representation of your company or brand — is it serious or fun, is it heavy or light, is it nostalgic, sci-fi or traditional, is it serif-like or sans serif-like, or is it glossy or matte. Match these questions against the company's impressions and perceptions that was unearthed in the first part of the branding process and see if there are specific traits that fit these values.

You logo should also not be confused with similar logos, especially those of your competitors. It should be a unique logo that is highly recognisable and easily associated with the company. For a start, sketch these ideas down based on what you have found out about the company. Develop these sketches further and create a

Rather than just a silhouette, the Rhino is designed to look more active. The rhino is shown to be charging rather than standing and doing nothing.

Designed by Estúdio Alice for Rotesma Indústria de Pré-Fabricados de Concreto branding (See page 062).

variation of basic graphical symbols and pictogram, or try combining these ideas to see how it could generate new ideas. Think of imagery that can have double meanings or hidden graphic elements, this could be a smart way to make the logo more interesting, and hence more memorable. The logo could also be developed further by turning a passive image to an active one.

For instance, the company or the designer have chosen to use an image of a rhino in their logo. The first reaction would be to create a silhouette of a rhino, keeping it as simple as possible so that it is easily recognisable and memorable. A silhouette with the rhino is passive — just standing there. In contrast, Estúdio Alice have made the rhino looked like it is charging forward, making it active (doing something) rather just standing there. In that way, it feels like the company is taking action rather than just waiting for things to happen.

For logotypes, do not use a typeface that is already strongly associated with other brands, for instance the font Klavika is strongly associated with Facebook, and the Bello script font used extensively in the old Airbnb logo. There are exceptions to this rule — typefaces that are used frequently — typefaces like Helvetica, Futura and Verdana are more generic and can work for most occasions. To enhance the look using common typefaces, make adaptations and slight adjustments to these existing typefaces — curve the edges, remove or add some elements, increase letter spacing and kerning, or combine it with graphic elements and illustrations — to own the

Examples of logotypes, where the typeface is adjusted slightly to create a unique logo.

Designed by StudioBrave for 42 Wills Street (See page 236).

Designed by Vide Infra for Level (See page 198).

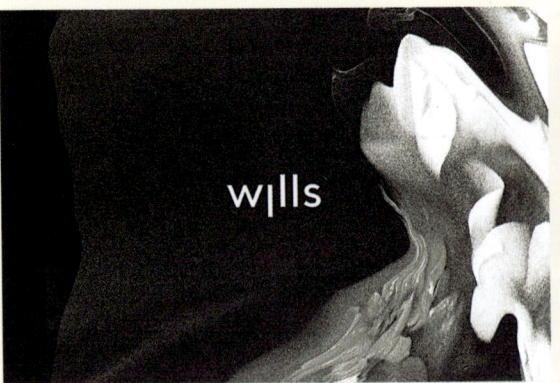

logo and to make it a better fit for the company's brand. Although we should not steal designs from other designers, it is also beneficial to see how other designers treat their typeface and adapt these smart tweaks to suit your company's needs.

When making logotypes, avoid using fonts that are gimmicky, especially free decorative fonts. Although free fonts sounds tempting to use (because it is free), they are probably being used in many poorly designed logos around the world. Besides, the typefaces are free for a reason — either it was used for testing, or they are not designed well. Also, do be wary when using "trendy" fonts. Trendy fonts generally go out of date quickly and a re-brand is required every few years. It is best to design a logo that is timeless, and it should last 10 – 20 years at the very least.

A unique logo that stands out from its competitors will reinforce your brand, and convey the company's values with the appropriate tone of voice. Make sure the typeface used in the logo matches the tone of voice of your company. A serious brand warrants a serious typeface, likewise, a kids company should have a typeface that is slightly light hearted and less serious typeface. Alternatively, think of the company as a person who is speaking to you, is he or she formal, friendly, serious, humorous or approachable? For example, a company specialising in homemade honey, that sells their products along a countryside road with a handwritten signboard, is quite likely to attract the attention of a passer-by looking for a rural experience. On the other hand, if your property agent advertises themselves using a handwritten signage, you might have doubts about their professionalism.

A logo should also be flexible and consistent. The logo should be able to work on multiple mediums: it should work well when it is enlarged (for billboards) or minimised (for namecards); it should work well in colour, and in black and white (for newspapers); it should work well in print and on screen. In most cases, logos tend to be as simple as possible to be flexible in its usage, as well as providing clarity and memorability to the public.

Last but not least, if the company have plans to expand further into the international market, be careful of cultural differences. With the world being interconnected and globalised, most companies will have to understand the symbolism that their logo might mean in other countries and cultures. For example, a "thumbs up" is generally considered a positive sign, a sign that means everything is good. However, in Sardina, Greece, and some Islamic countries, it is a rude sign with sexual connotations. Presenting a logo with a "thumbs up" symbol for a Greek restaurant is an absolute no go.

2.2.3   Extension

Even the best general needs the best commanders and soldiers to be able to win a war. Likewise, just having a great logo does not translate to success for your branding. A brand identity system is usually required so that all aspects of the company is consistent and cohesive. A logo should be backed up by a brand identity system with an appropriate tone of voice in all aspects of operations within the company. Based on the research done in the first part of the branding process, analyse logos of your competitors and companies who the brand aspires to be. Pinpoint what works and what doesn't — from the colour choices and palettes, the typefaces, imagery and photography, as well as the tone of the voice of the tag-lines used. Do remember that in this age of social media and seamless connectivity, the tone of voice in your tweets and posts could affect how customers perceive you instantaneously.

A brand identity system is an extension of the logo, that includes stationery (namecards, letterhead, envelopes), marketing collateral (brochures, mailers, advertisements), online presence (website, mobile site, social media), as well as any other mediums and materials that are necessary. It could also extend to objects such as safety helmets, bollards and even construction tools, for companies in the construction and property industry. A brand identity system also extends more than just objects, it also includes other elements that make up the company's visual language, for example, the copy-writing style and the photographic approach. More importantly, all these extensions should be cohesive and recognisable even without the presence of a logo.

There are several ways to create a consistent and cohesive brand identity system — the colour and the typeface.

One of the fastest and easiest way to create a distinct identity is to own a colour for your brand. However, at times using one colour isn't sufficient to create materials for the company, and most brands use a combination of one to three colours, with one colour being the primary colour. Consider developing a palette of complementary secondary colours to provide flexibility. Although the branding will focus on one colour, your secondary colours can provide greater impetus into your marketing and promotional materials. Depending on the needs of the company visual identity, shades and tints of the primary colour can also be used to bring out a more extensive palette of colours.

Strata is a landmark commercial property development in Staines upon Thames. The image on the right showcase the complete brand identity system, displaying how the logo and its accompanying graphic elements are being applied on the brochures and box-sets, down to how the floor-plans are being designed.

Designed by Blast Design Ltd for Strata (See page 170).

A chart illustrating how different colours are being perceived by the public, and what these colours say about your business.

Source:
entrepreneur.com
thelogofactory.com
inc.com
logodesignworks.com

Energy Finance Airlines Automobile Food Agriculture Tech Medical Fashion Property

WARM COLOURS
Warm colours are often associated with energy.

COOL COLOURS
Cool colours are often associated with calm and security.

RED
Evokes a passionate and visceral response, a colour that increases the heart rate.

Aggressive / Energetic
Provocative / Attention-grabbing

PURPLE
Sophisticated yet mysterious, a colour of royalty and elegance.

Royal / Sophistication
Nostalgia / Mystery / Spirituality

YELLOW
Communicates hope and optimism, a colour that stimulates creativity and energy.

Positivity / Light / Warmth
Motivation / Creativity

BLUE
Most popular choice for a brand, a colour that puts people at ease.

Trustworthy / Dependable
Secure / Responsible

ORANGE
Combining the cheerfulness of yellow with the energy and boldness of red, a colour of excitement.

Vitality / Fun
Playful / Cheerfulness

GREEN
Calm, freshness and healthy, deeper tones are associated with affluence.

Wealth / Health
Prestige / Serenity

BLACK
Classic and sophisticated, works well with luxury and expensive products.

Prestige / Value
Timelessness / Sophistication

BROWN
Simplicity, strength and durability, use caution as it reminds people of dirt.

Earth-like / Natural
Simplistic / Durable

WHITE
Purity and cleanliness, popular choice for healthcare and children businesses

Pure / Noble
Clean / Soft

While a logo should not have too many typefaces, a brand identity system requires a strong family of typefaces for different occasions. It is best to select a typeface that will be a strong match to your logo and other visual elements. In addition, a secondary typeface could also be required to provide contrast and hierarchy in your marketing and promotional materials. When matching fonts, do that note that it should belong to different classes to create contrast, for instance, a sans serif font would work well with a serif font, and a decorative font could work well with a san serif font and so on. A family of typeface is great, but do considering engaging a graphic designer or type designer to create a font for the logo and its accompanying tag-lines and headers. In this way, your brand would truly stand out and be unique. Companies like Uniqlo, Heineken and Walt Disney created their own unique fonts for their brands and are exclusively used by them.

Understanding what serifs and sans serif fonts can do to your brand is important to creating eye catching logos.

On the right, see the difference between san serifs, serifs, decorative and script typefaces. The grouping of fonts are commonly referred to as font classifications.

Even width

Strokes that have variable widths

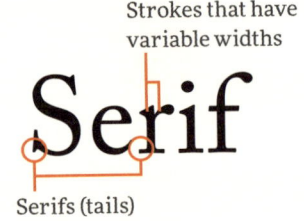

No serif (tails)

Serifs (tails)

Sans Serif is best used for websites and flat designs, and for creating serious headers or copy.

Serif is best used for newspapers and novels, with a strong and bold personality.

Novelty typeface, creating many varying moods, from classy to elegant to light hearted.

Cursive or handwritten typeface that are ornamental. Best used for a soft and classic feel.

Creating contrast is important to making your logo more eye catching, rather than a single font family being used.

Thin and lowercase letters

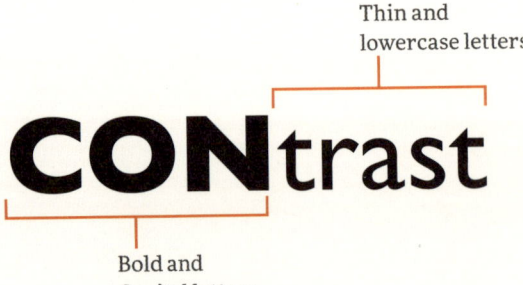

Bold and Capital letters

Create contrast by using uppercase and lowercase letters, making it unique and eye catching.

Different colours can be used to create contrast so that it is more attractive.

To show contrast, we pair different fonts of the same classifications (Sans Serif logotype, with another Sans Serif font for the tag-line), or we pair the same fonts for both the logotype and the tag-line (Same font, but difference in colour and size).

Sans Serif
Same font but lighter colour

Serif
Same font but lighter colour

On most occasions, the safest way is the pair fonts from different font classifications. E.g. Sans Serif logotype with a serif tag-line, a decorative logotype with a sans serif tag-line, etc.

Sans Serif
Serif to show contrast

Serif
Sans Serif to show contrast

2.2.4 Brand Guide

After discovering what the company is and what the company wants, and after designing the logo and the brand identity, what is next? Usually, the designer would hand over the logos and brand identity to the company, and the next thing you know, the logo is plastered all over the place. Often, the branding project has changed so much that designers and agencies no longer wanted it in their portfolio, and one way to ensure a smooth handing over is to produce a brand guide.

Increasingly, more and more designers have provided a brand guidelines document to illustrate how their logos and branding should look like. This serves as a guide for consistency and cohesiveness when the logo and its accompanying elements are being used. Usually, the document will include details on the minimum dimensions of the logo before it turns illegible, the versions of the logos and which should be used on different occasions (black and white or colour, horizontal or vertical) and how these logos should be treated (squashed, stretched or muted colours). It will also include details on the colour swatches (CMYK, RGB or Pantone) as well as primary and secondary fonts.

What to include in your brand guidelines?

1. Brand Overview
▶   A short description of your vision and the brand essence. Provide clear statements and keywords that people should keep in mind while designing or creating content for marketing collateral.

2. Logo
▶   The logo is an important part of the brand identity and it should be taken seriously. Provide logo variations and state the minimum size of the logo before it becomes non-legible.

An example of a brand guide, showing the types of typefaces used, the colour palettes and combinations, what you should not do when using the logo, as well as examples of the logo's application on other stationery.

Brand guide, designed by Jazz Design for Francelino Imóveis (See page 162).

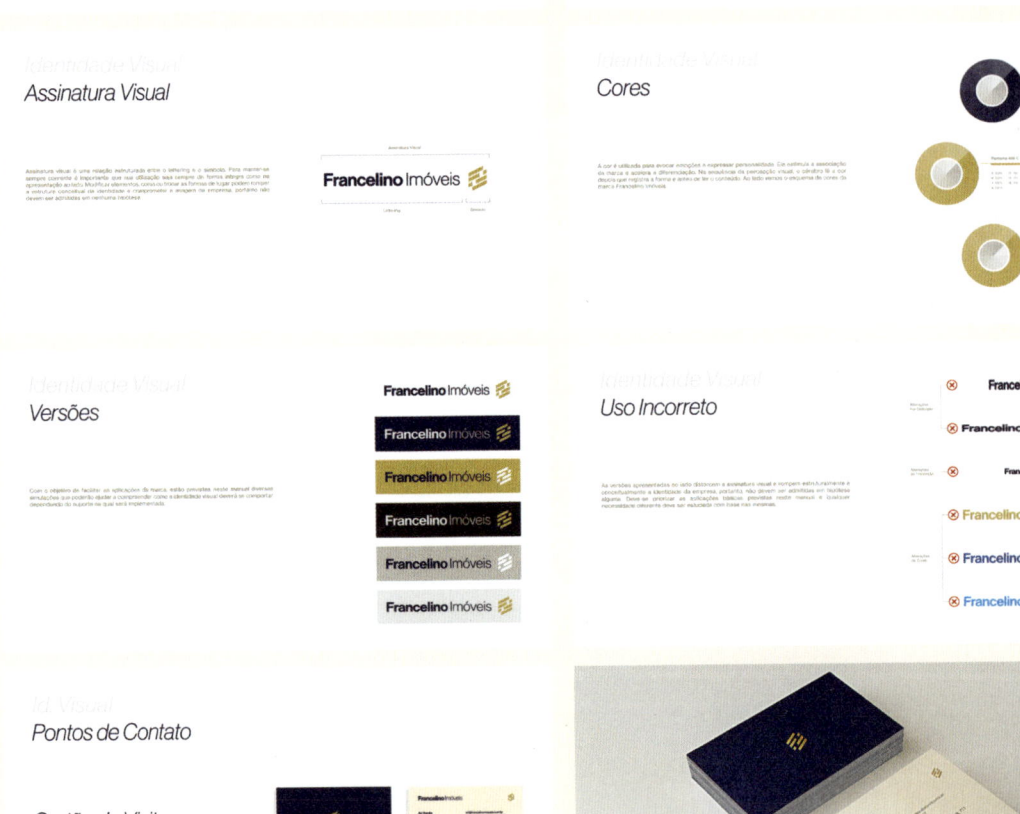

- ▶ Logo variations could involve the creation of horizontal and vertical logos, as well as logos that only show the typeface or the symbol, or both.
- ▶ Provide other logo variations and specify what colours are allowed.
- ▶ In addition, show examples of what to do and what not to do when using the logo. For example, some logos are not allowed to be rotated, add on effects like shadows and emboss, outlined, stretched, or even be replaced with other graphics.
- ▶ The guideline of logos should also consider the spacing between logos and other elements. Most non-designers do not consider the use of white space as an important aspect of presenting your brands. Often, brand guidelines state the space in numbers, but rather than doing that, it would be easier to understand if you take an element of the logo and explain it based on the height of the element.

3. Colour Palettes
- ▶ Include colour palettes in CMYK, RGB or Pantone for use on their respective mediums.
- ▶ Also specify primary and secondary colours and explain the use and situations where these colours can be used.

4. Typefaces
- ▶ Identify and define the typeface to use for your brand identity. This could include the size of the typeface, the family of typeface used on the headers, sub header and body copy, the colours, and the complementary typefaces.
- ▶ Do remember to provide alternatives for web, as there are limitations to some fonts online and for screen usage.

5. Copy-writing
- ▶ Copy-writing is an important aspect of branding, especially for companies who are required to write their own copy.
- ▶ Might seem unnecessary, but the brain registers the company's messages subconsciously and bad copy-writing could ruin the reading flow.
- ▶ For example, you can create rules such as using words when writing "one" to "nine," and indicating the use of numerals from 10 onwards, or providing a clear way to write dates (Wednesday 2 March 2016 and 2.3.2016), etc.

6. Imagery
- ▶ Establish a particular style for your photographs and a particular tone and style for your images and illustrations.
- ▶ Explain why the style is preferred and why it will work for them.
- ▶ For example, some banks might not want their photograph images to have people staring straight at the viewer, this "stare" might be too intimidating and the banks would prefer not to have it that way.

Compilet these information in a well-designed booklet for the company. Make the branding guidelines easy to access so that everyone, both internally and externally, can use it as a guide. While considering the ease of access, design it so that it is printable and viewable on the computer screen. Often, people get annoyed by its enormous size that they do not read it. Also, make two versions of the brand guidelines: a comprehensive guide detailing all the aspects of the branding, and a pocket guide where people can use it for quick reference.

Another example of a brand guide. This brand guide also details the use of imagery for promotional purposes as well as the corporate presentation layout. The brand guide also go to the extend of showing how the email signature should look like, to maintain consistency across multiple mediums.

Designed by Vide Infra for Futuris (See page 184).

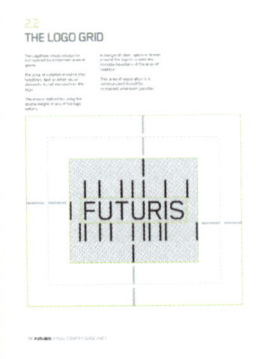
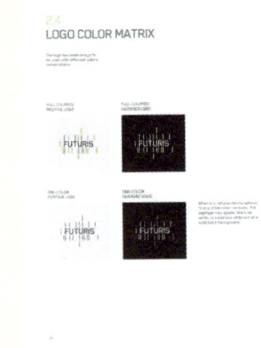

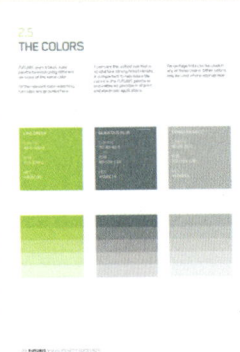

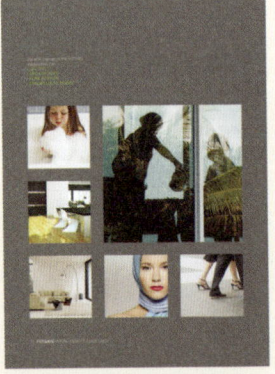

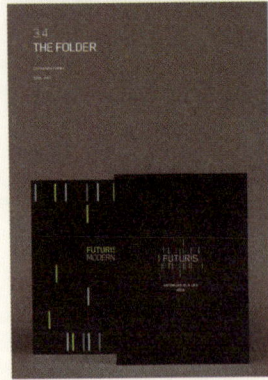
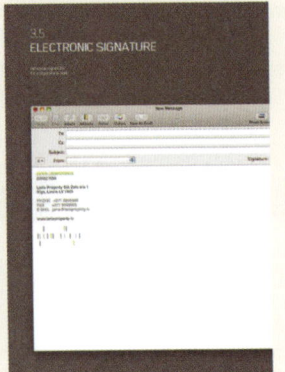

2.3 Renewal

The last part of the branding process, albeit a never ending one. The company and the branding agency should refine, tweak and reject where necessary. Look through the entire process and use the research that was conducted in the discovery stage. This is the time to be objective: removed or review any process or elements that do not match the objectives of the branding process. If you are unhappy with the outcome, it is the time to sort it out before your brand is being announced to the world.

3 CONCLUSION

Remember that scrimping on the design and printing can end up costing your brand — financially as well as psychologically — if it is not up to standard. First impressions count and the failure to make a good impression could result in more damage control measures over the years. If you do require assistance, it would be best to ask around for recommendations. Talk to designers and agencies and chose one who understands your target audience and have values that you share. Often, designers are lateral thinkers and have their areas of expertise, an industrial designer helping you to code a website is a recipe for disaster.

Last but not least, we hope that this guide could help your brand stand out from your competitors and improve your business. The process of branding is never easy, but it is totally worthwhile and beneficial. Good luck building brands!

BUILDERS

Companion

COMPANION

CREDITS

Studio — **Sabbath Visuals**
Creative Director — **Jorge Zamonsett**

DESCRIPTION

Companion is a real estate company made up of a small but expertise team. Companion claims and supports human warmth and professionalism in order to offer wonderful guidance in the search for patrimony or properties.

The name is developed from a direct connection — Comitis — or a person of trust.

A modern coat of arms consist of a lion: an element that represents strength, respect and conviction. A secondary element — a key — represents equilibrium and loyalty.

As part of the visual language, we develop a nexus of four values as pillars for the company that is represented in Pantone colours.

The values function as the catalyst and represent the efforts which make the company profile unique, elegant, innovative and reliable.

INVERSIONES CAPITAL

CREDITS

Studio — **FUTURA**
Photographer — **Caroga**

DESCRIPTION

"Inversiones Capital" is a real estate development company located in San Pedro Garza García. Even though they seem to be a new company, they already existed before with another name, so the most important value that we needed to communicate with the branding was their experience.

We worked on an emblem inspired on the figure of a lighthouse, making reference to the fact that this Company is a guide for their clients and partners not only by supporting them with an important background and knowledge but also with personal advice through each and every project.

The typographical selection counteract with the illustration technique and colour selection that screams luxury, it reminds us that this is also a young company with a strong vision to the future.

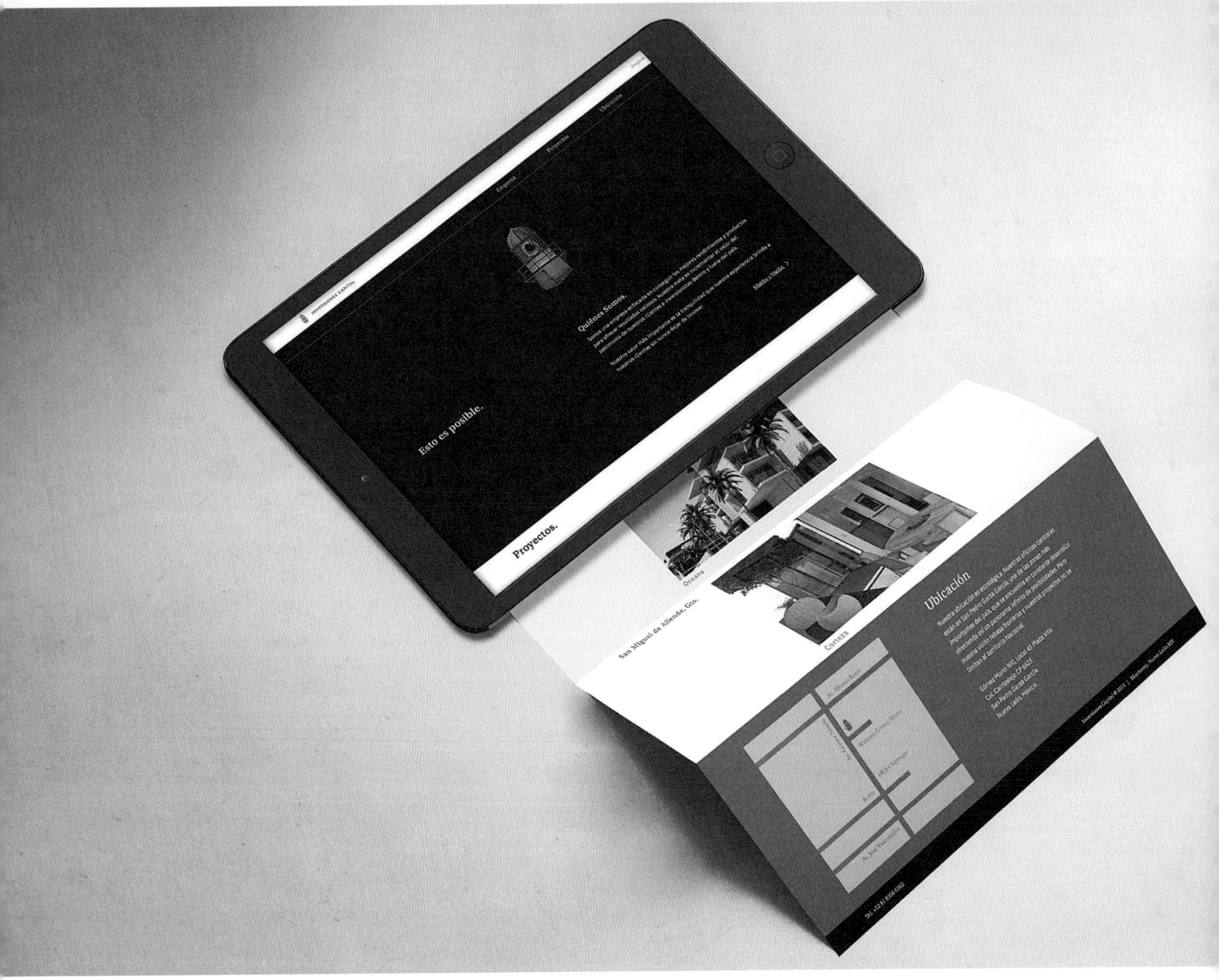

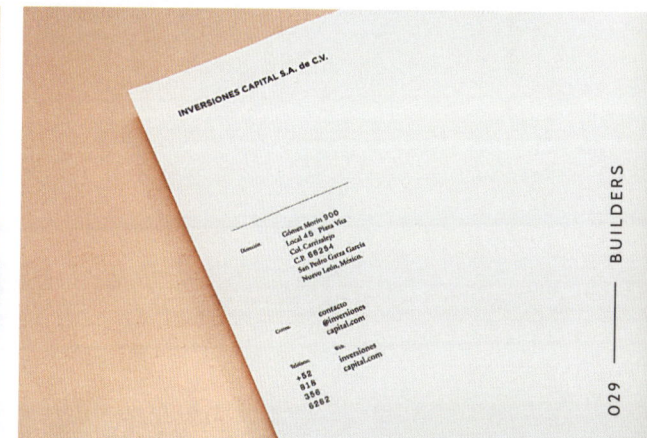

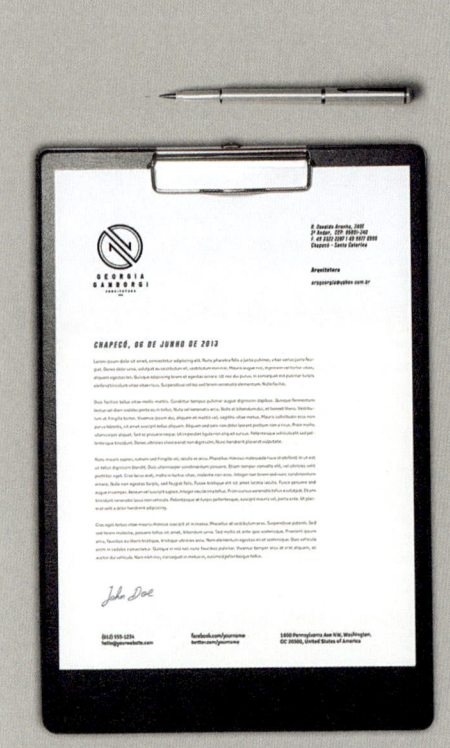

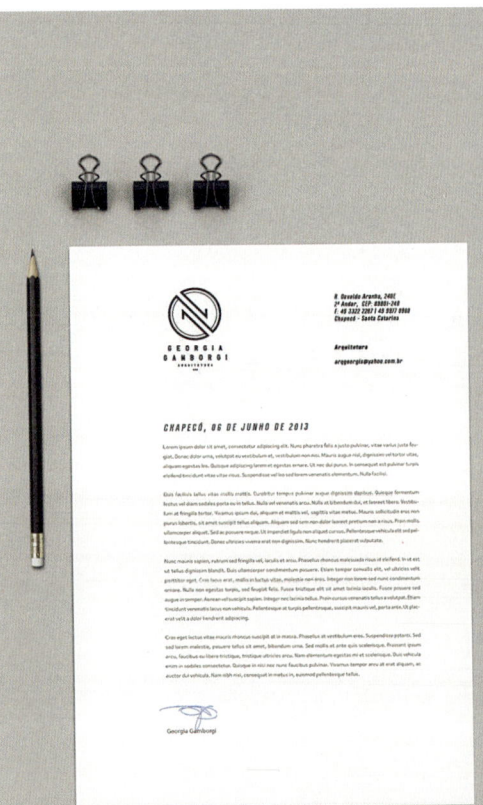

GEORGIA GAMBORGI ARQUITETURA

CREDITS

Studio — **Estúdio Alice**

DESCRIPTION

The influences came from a survey of the architects who are the soul of the office. We proposed a simple challenge that each gather one item from different cultural segments that are part of their everyday life — from architecture references, design, art, movies, music, etc. With this creative process, we had a better context of this intense universe, understanding that the branding needed a straight visual as an architectural firm — explosive and creative as the image of products and services.

RESIDÊNCIA CONTEMPORÂNEA

GEORGIA GAMBORGI
ARQUITETURA

Lorem ipsum dolor sit amet, vel ne illud munere repudiandae, an reque adolescens eos, te est euripidis theophrastus. Cu vel quodsi adolescens mnesarchum, cu quas exerci neglegentur vim. Has no inani cotidieque definitionem, fastidii rationibus his te. At solet graece everti nec. Nam no mundi reprimique, aliquid omittam pri et. Pro volumus phaedrum repudiandae ut, ei his altera mucius eruditi. Eirmod latine constituto mea ei, eius choro saperet nec id, eam ex eirmod euripidis. Malis labore no vis, sonet civibus laboramus pro ad. Sit cu soluta repudiandae, mazim laboramus constituam est cu, pri nonumes iudicabit tincidunt ad. Sea stet reque placerat te, nam an saepe postea vidisse.

GEORGIA GAMBORGI
ARQUITETURA

GEORGIA GAMBORGI

Lorem ipsum dolor sit amet, vel ne illud munere repudiandae, an reque adolescens eos, te est euripidis theophrastus. Cu vel quodsi adolescens mnesarchum, cu quas exerci neglegentur vim. Has no inani cotidieque definitionem, fastidii rationibus his te. At solet graece everti nec. Nam no mundi reprimique, aliquid omittam pri et. Pro volumus phaedrum repudiandae ut, ei his altera mucius eruditi. Eirmod latine constituto mea ei, eius choro saperet nec id, eam ex eirmod euripidis. Malis labore no vis, sonet civibus laboramus pro ad. Sit cu soluta repudiandae, mazim laboramus constituam est cu, pri nonumes iudicabit tincidunt ad. Sea stet reque placerat te, nam an saepe postea vidisse.

CATHERINE
JACOMEN
PAINTING

CATHERINE
JACOMEN
PAINTING

CREDITS

Studio — **Core Agency CA**
Creative Direction + Design — **Core Agency CA**

DESCRIPTION

Brand identity for Catherine Jacomen, a painter and a decorator based in Stratford, Canada.

KSAB

CREDITS

Studio — **Nhomada**
Creative Director — **Diego Leyva**
Photographer — **Ana Georgina**
Graphic Designer — **Diego Leyva**

DESCRIPTION

KSAB offers solutions for sustainable problems in Mexico. Solar energy green construction and ecological solutions are the ground of Ksab's vision. A vibrant spectrum of colours was selected in order to create a strong visual language, creating a cheerful brand expression.

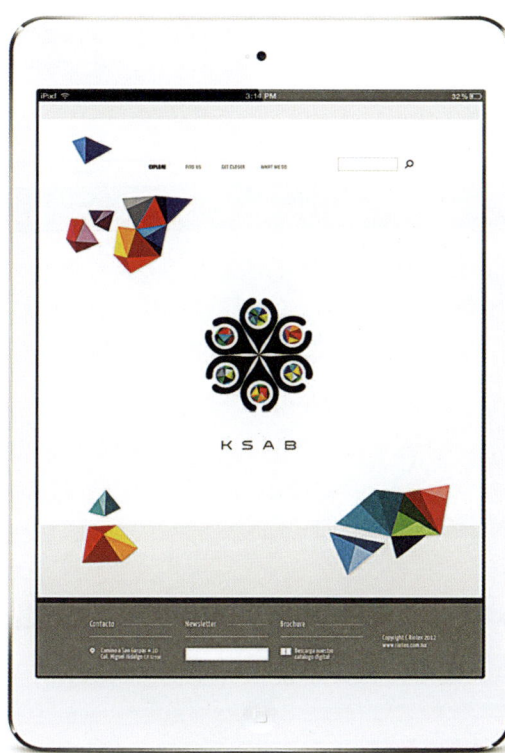

ARCH
RESIDENTIAL

Arch Residential
12 Trinity Street
London
SE1 1DB

www.archresidential.co.uk
T. 0207 111 4983
F. 0207 183 3513
E. info@archresidential.com

ARCH RESIDENTIAL

CREDITS

Studio — **DM Workroom Ltd**
Art Director — **Denis Mallet**
Designer — **Denis Mallet**
Photographer — **Denis Mallet**

DESCRIPTION

Arch Residential is a lettings focussed agency based in London. As the main constitutive logo element we created an iconic symbol from the first letter of the company name. We have decided to create a spiral around the letter A to enhance the territory expansion concept but also to create a structural and architectural depth, enriching the arching construction. The idea was to dissimulate an architecture shape, a hall with arches that leads to the central logotype, which represent a door and the access to property. It is also possible to perceive stairs leading to a door.

Arch Residential is a real estate agency and being able to dissimulate an architectural optical sensation was a difficult task that resulted from a well proportioned graphical construction. The effort regarding the proportions was mainly on the thickness of the line that constitutes the logo. This line can also expend as an infinite spiral on a blue colourful background. Choosing the right stroke thickness on each document is primordial to get a good balance between the blue and bronze metal colours. Such type of repetitive lines and colour contrast can easily become annoying for the eyes and often creates kinetic optical vibrations that can be extremely irritating and tiring for the viewer. Looking for the right proportion was really important to avoid any severe optical vibrations or moiré.

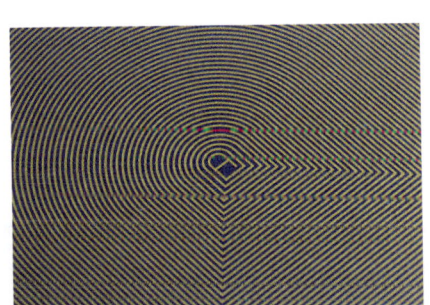

ARCH
RESIDENTIAL

Arch Residential
12 Trinity Street
London
SE1 1DB

www.archresidential.co.uk
T. 0207 112 4983
F. 0207 183 3113
E. info@archresidential.co.uk

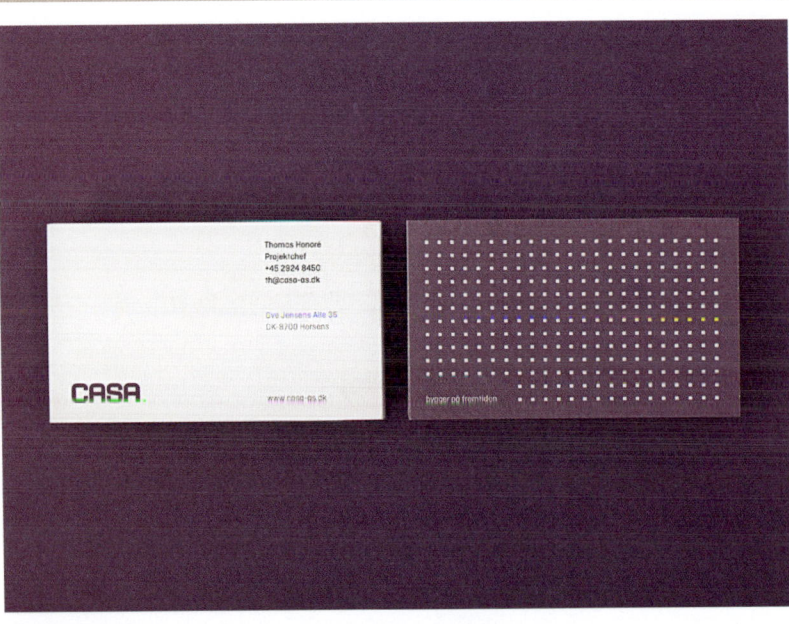

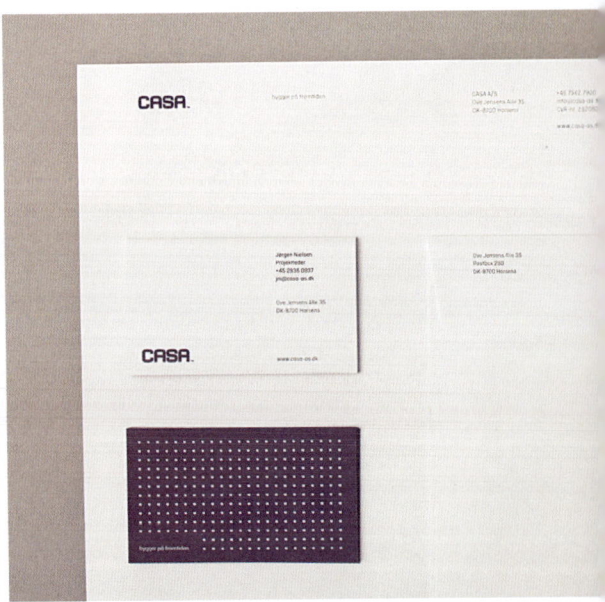

CASA™

CASA

CREDITS

Studio — **Ineo Designlab®**
Account director — **Søren Herold**
Creative director — **Peter Christensen**

DESCRIPTION

Identity design for Danish construction and property development company CASA.

CASA is a highly successful Danish construction and property development company that develops, builds and rents residential, commercial and public buildings throughout Denmark.

The Hugo
205 – 207 Ballarat Rd, Footscray

FAYMUS

Fraz Tanvir
Vanguard & Developer

TITLE

FAYMUS

CREDITS

Studio — **StudioBrave**
Creative Director — **Tim Sutherland**
Designer — **Mike Nguyen**
Illustrator — **Mike Nguyen**

DESCRIPTION

We have been fortunate to work with Faymus
across multiple platforms. While developing
brand marketing campaigns for multi-residential
projects we have also created their own brand
identity while building their brand profile.

They are truly a property development company
with a difference. Their principles are bound in
interlinking tradition and innovation to create
natural, beautiful and sustainable living.

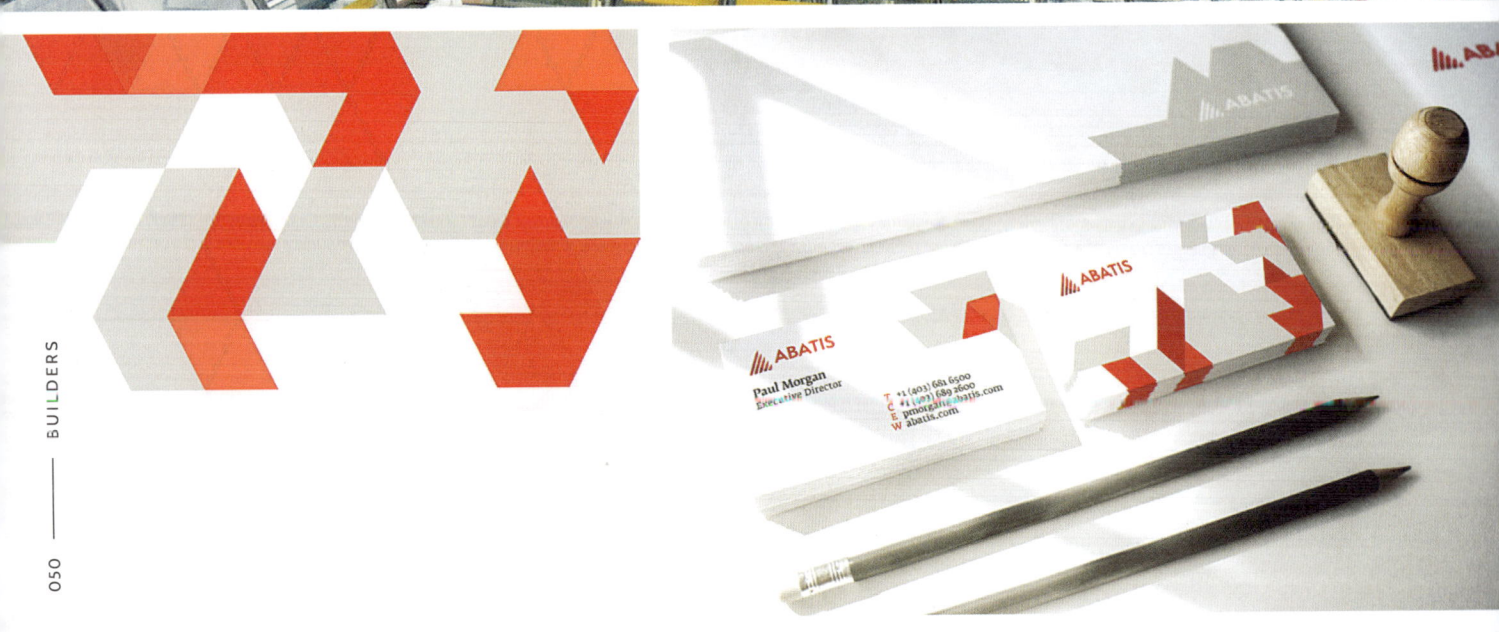

TITLE

ABATIS
BUILDERS

CREDITS

Creative Director — **Core Agency CA**
Designer — **Core Agency CA**

DESCRIPTION

Abatis Builders are an organisation with a primary focus on professional residential and commercial contracting and design. Abatis transform visions into reality to produce a beautifully functional space within a given budget providing exceptional value and service.

BOURKE STUDIO

CREDITS

Studio — **makebardo**

DESCRIPTION

makebardo was engaged to create the visual identity for Bourke Studio, Architectural Planners based in Melbourne, Australia. In this case we felt that effective use of the colour palette and shapes contributed to a distinctive brand identity.

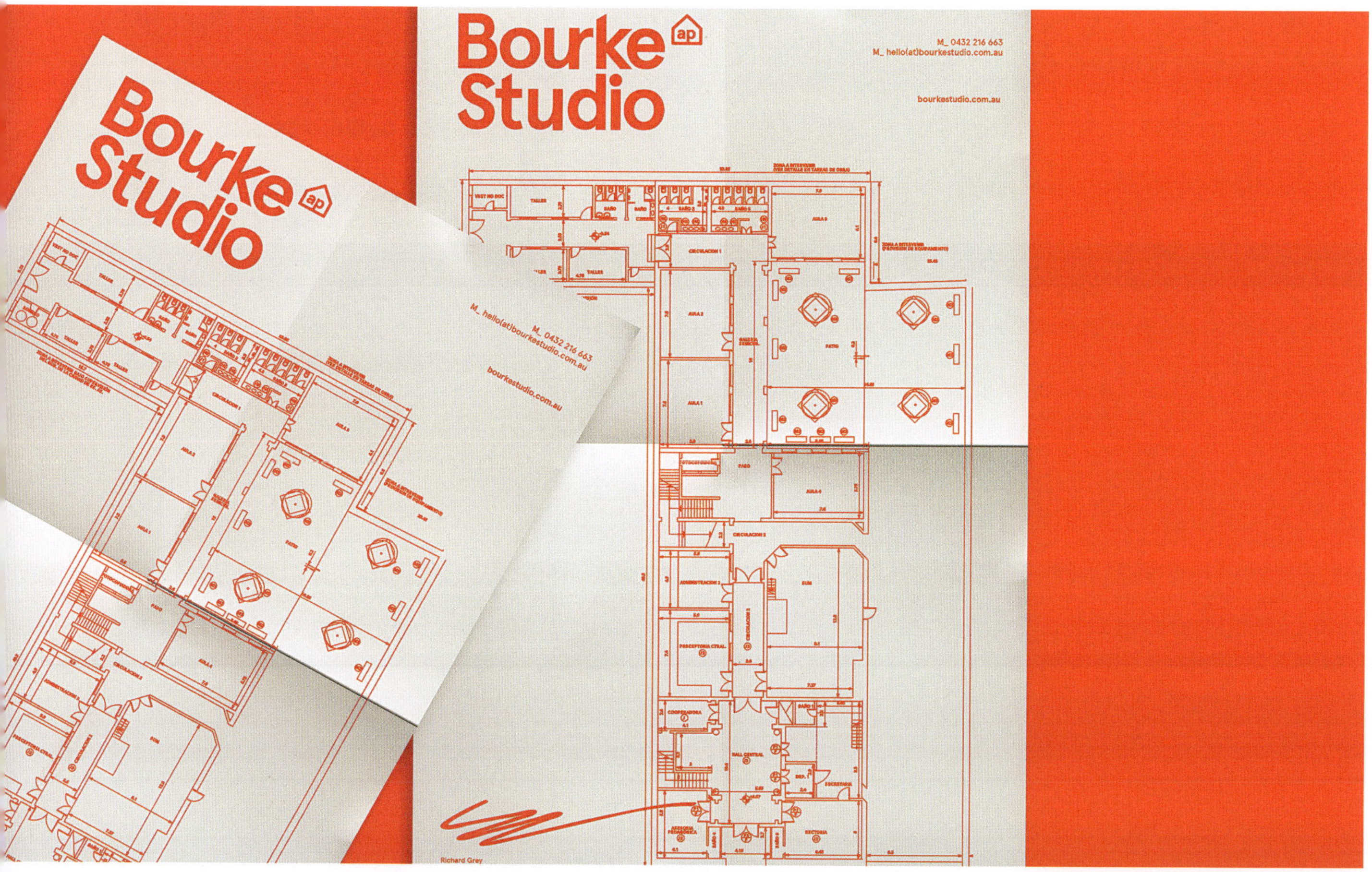

TITLE

ECOMS

CREDITS

Studio — **Actual Studio**
Graphic Designer — **Jonathan Ford**

DESCRIPTION

ECOMS is a new Cardiff-based company that provide project management services for the construction industry. We designed a logo and identity for the new venture, as well as bespoke, responsive website to showcase their services and capabilities.

ECOMS, 2 Deere Close, Cardiff, CF5 4NU
...ited.co.uk • E: mark@ecomlimited.co.uk • T: 07414 733 193
...Management Services Limited • Company registration number: 8627085

TITLE

HOUSING

CREDITS

Studio — **Moving Brands**

DESCRIPTION

Housing was born out of its founders' own struggle to find a home. In revolutionising the local real estate market, Housing has grown, in under three years, from a small team in Mumbai to 1,500 employees in 45 cities across India. Housing's data science labs and innovative product design is attracting the attention of the global tech scene, as well as investors; Softbank has led the $121M investment in the company to date.

After seeing Moving Brands successful partnerships with other global tech businesses, we were invited by Housing to join their mission as their lead creative partner. We have worked side by side for a year to create a brand narrative, identity system and brand language; to define strategic cross-media communications plans; to support the experience design of the app and design and build of the microsite; and to create the launch campaign communication assets that reveal the brand.

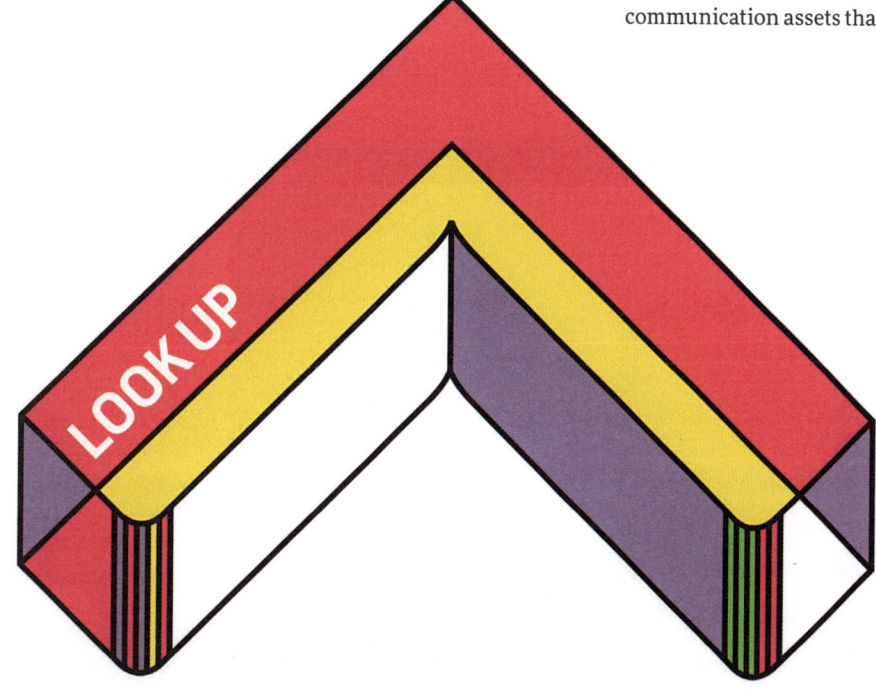

SALLY DERNIE

CREDITS

Studio — **Phage Ltd.**
Designer & Creative Director — **Natasha Zlobec, Danny Brooks**

DESCRIPTION

Phage worked with Sally Dernie on a complete brand transition to reflect a shift in direction for her high-end interior design practice and a maturing of her target audience. The new corporate identity, including logo, communication materials, and website, were all designed to create a luxurious, and eclectic feel, with high-end finishes, such as the die-cut scalloped edge on the company stationery, combining with metallic inks and warm colours to reinforce the richness of the brand.

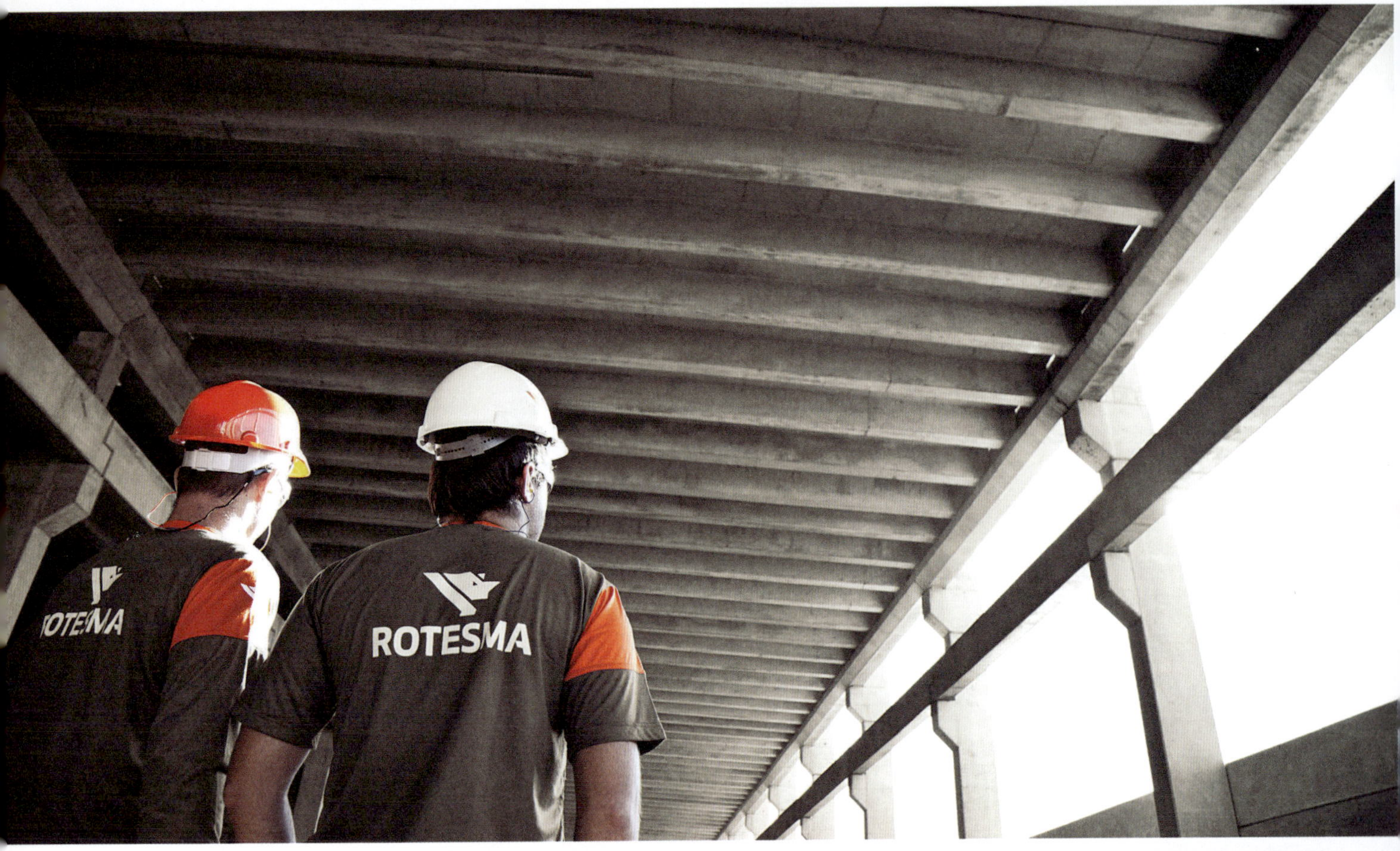
ROTESMA
ROTESMA

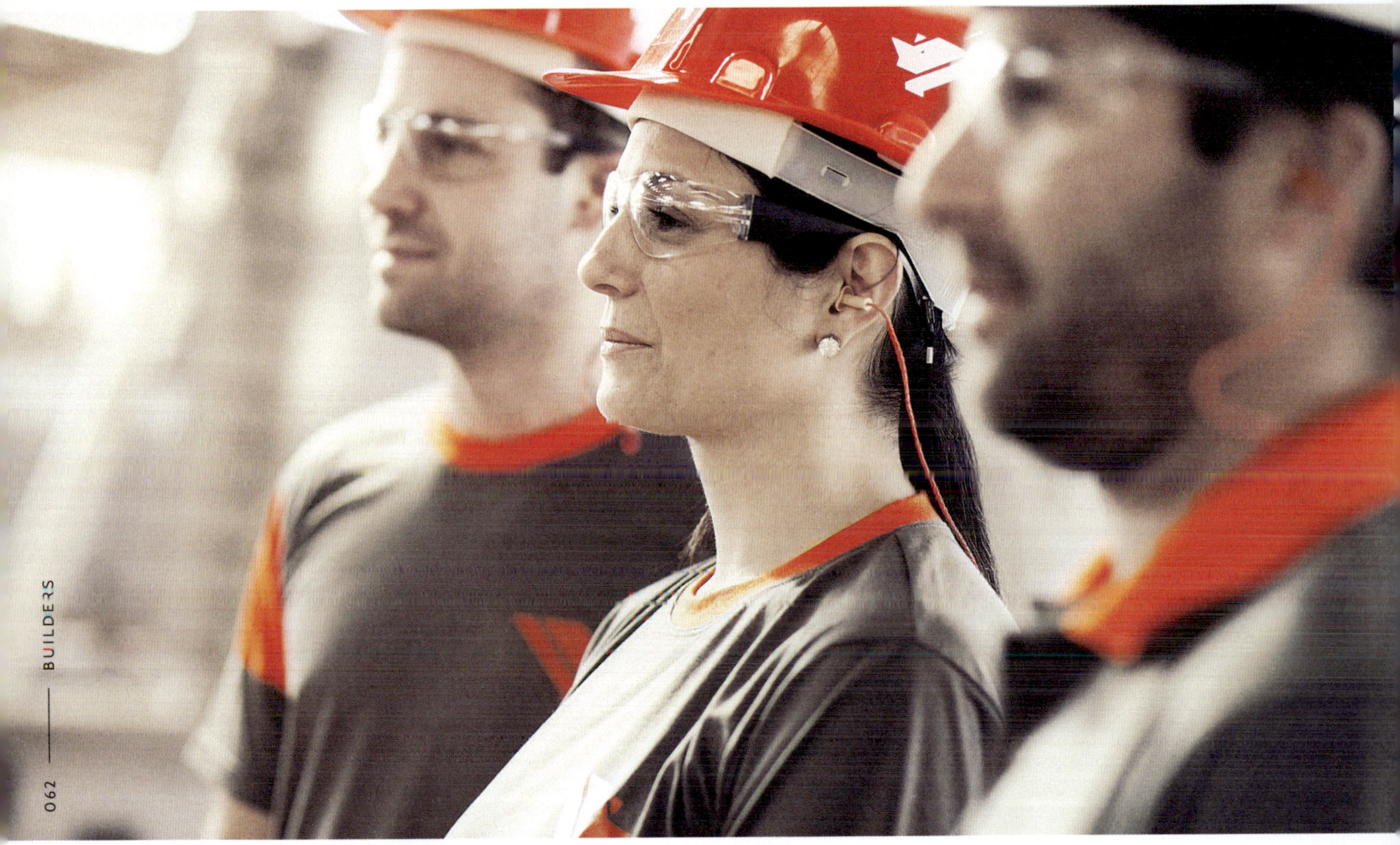

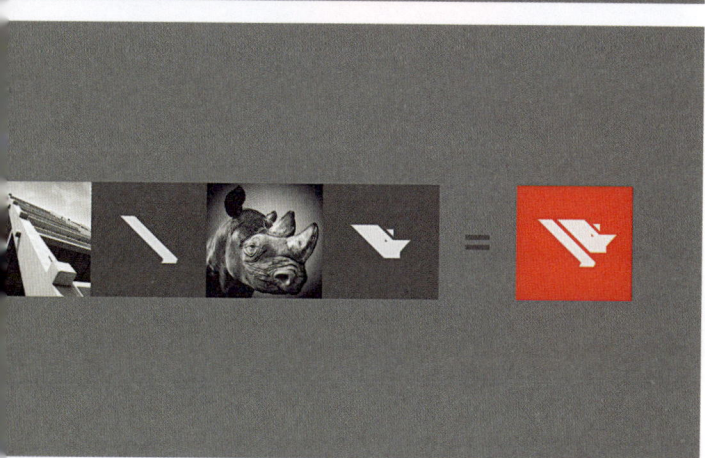

ROTESMA INDÚSTRIA DE PRÉ-FABRICADOS DE CONCRETO

CREDITS

Studio — **Estúdio Alice**

DESCRIPTION

The Rotesma is a precast industry that has been operating for over 30 years in the segment. We needed to understand this process of transition and align the communication with these new strategy of nation market insertion. We were consulted to think the visual of a catalogue, but the symptoms already pointed to a necessity of a re-branding and strategic repositioning. And the diagnosis was immediate. The brand did not had an icon, the colours used were very cold and had no differences in relation to the segment.

We set out to identify the needs with the challenge to reinvigorate the brand. When talking to the business marketing people, we came to the conclusion that the aggregate value of an animal for the mark would be positive. The fact that the company works with heavy elements and animal power would add value to imposing differently. We thought of animals that have these characteristics and to represent the strength and determination. We arrived at the rhino concept, seen as an animal with the largest number of necessary attributes.

But we needed to turn it into an icon, which translates into imposing manner without compromising the brand communication. The icon was built when we thought of ways to relate the design with prefabricated elements. What you see on the re-brand is the subjectivity of the animal that has become more present in the communication, but mainly it reinforced the concrete essence of the icon. The orange tone added in the material achieved prominence in the industrial park mainly in uniform and internal communication of the company.

TITLE

PATERNA

CREDITS

Studio — **FUTURA**
Photographer — **MarcomásChuy**

DESCRIPTION

Paterna is our favorite carpenter workshop. Our goal: Optimize each resource without wasting, this is the reason of the typeface that allows simple applications in any space.

A family project that became professional.

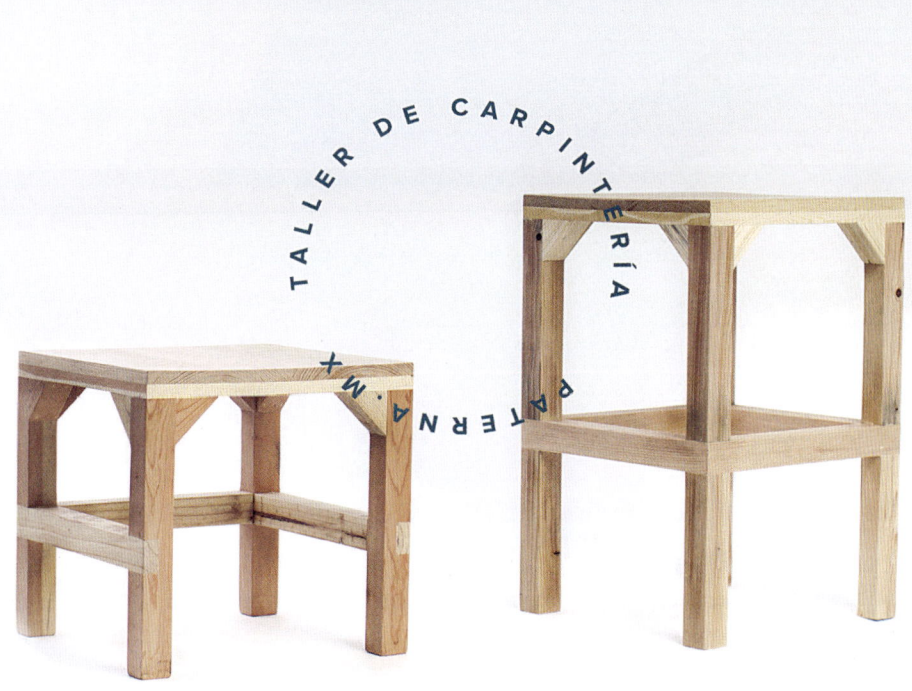

BUILDERS

LOL

Studio — **FUTURA**
Photographer — **Caroga**

DESCRIPTION

LOL is an architectural firm obsessed on finding beauty on the things they do. They apply in every project their concept of "Artistecture" — the action of turning design and architecture into art.

Our branding solution was based on that principle. We develop the stationer,y seeking that every part could reach more value beyond its function turning them into objects of desire. Subtle details of colour, printing finishes that add tactile sensations to the experience and a logotype that finds clearness in the chaos. They final result is a walk through a white gallery full of little surprises.

Jill Scholes
interior design

Jill Scholes
interior design

Jill Scholes
interior design

e course of a project and
e of the co-founders and
anding, print, and web
nts to communicate

not be tailored for
nse we have
o short-run
functional,
uniquely

e or
your

JILL SCHOLES INTERIOR DESIGN

CREDITS

Studio — **Phage Ltd.**
Designer & Creative Director — **Natasha Zlobec, Danny Brooks**

DESCRIPTION

Established design practice Jill Scholes Interior Design approached Phage to update their graphic identity so that it more accurately reflected their design style and personality.

Maintaining the essence of the original brand, we refreshed the logo, introducing colour and secondary brand elements to soften the look and nicely tie everything together. Multi-functional print solutions that could be tailored in-house cost-effectively delivered brand consistency throughout the practice's entire client process.

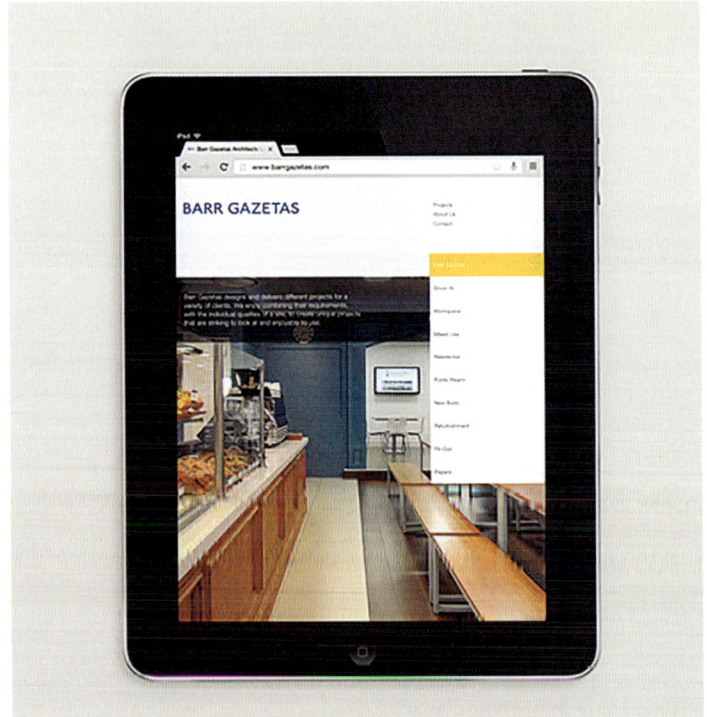
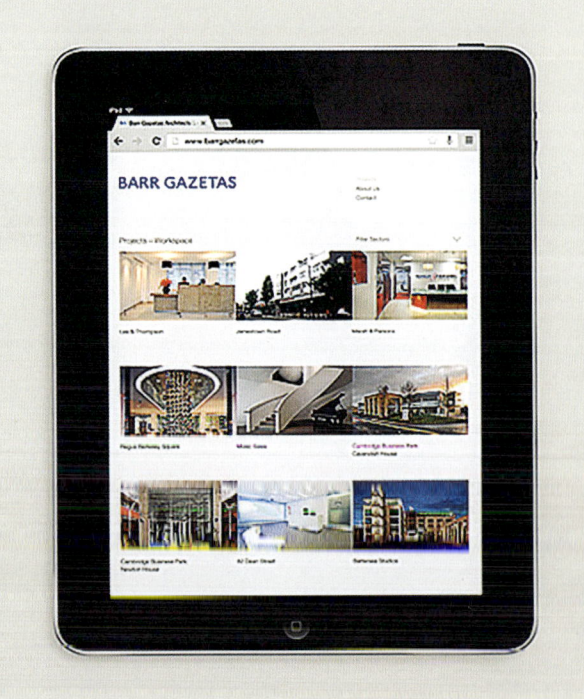

BARR GAZETAS

CREDITS

Studio — **Phage Ltd.**
Designer & Creative Director — **Natasha Zlobec, Danny Brooks**

DESCRIPTION

Barr Gazetas are an established London architectural practice, whose 20 year old visual identity needed a bit of a lift. Phage were brought in to consult on the update, leading to a complete refreshing of the company logo and corporate colour palette, the design of new presentation materials and in-house guidelines, and a new website that better showcased their portfolio of work.

With such an understated identity, the attention to detail had to be 100% spot-on. The new business cards demonstrate that perfectly; foil-blocked in the company's brand blue, with a sheet of blue board triplexed between two white sheets to give just a flash of matching colour along the edge.

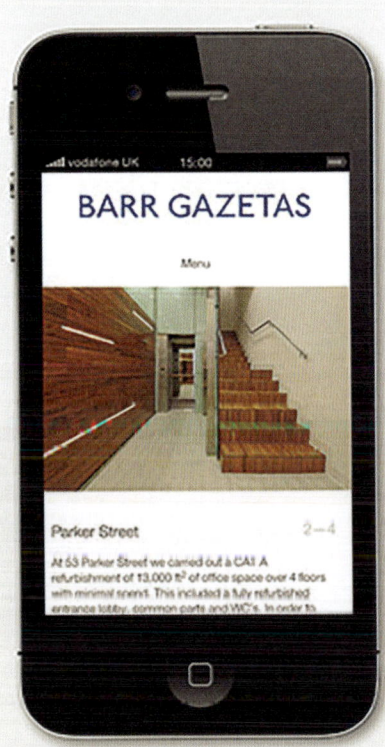 

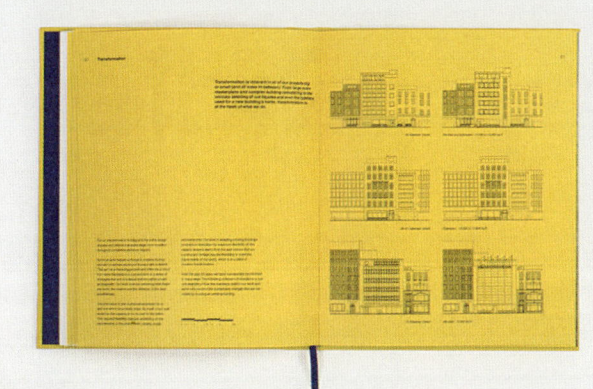

Australian firm Ridley is at the forefront of some of the biggest developments in the architecture and construction industry. They act as a central hub for architects, builders and developers, making sure everything comes together just so. Ridley has pioneered the use of Virtual Design Construction, the digital revolution that is reshaping the industry. By attaching live data to every part of a 3D model, they're able to gain an overview of the entire building process from start to finish. The identity plays on this idea of seeing the bigger picture, using individual elements, partial views and cropping to create the whole. The use of human-centric data helps everyone who engages with the brand better understand the company, projects and the people who work there.

Ridley Ridley
Ridley Ridley Ridley
Ridley Ridley

BUILDERS

076

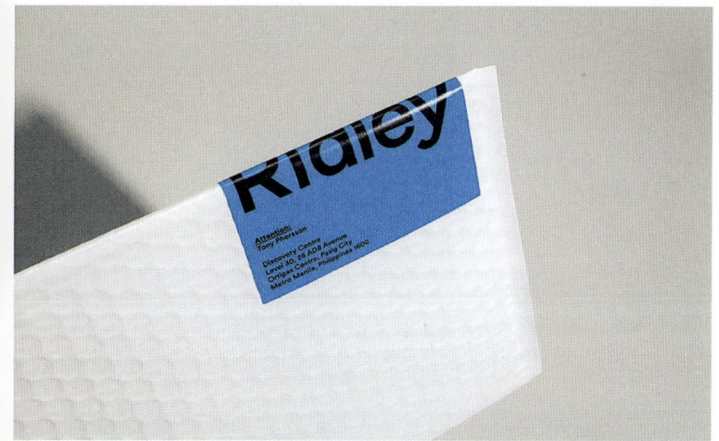

RIDLEY

CREDITS

Studio — **RE: Sydney**
Creative Director — **Jason Little**
Design Director — **Ryan Atkinson**
Designer — **Alex Creamer, Louise Elliott**
Strategist — **Gareth Stewart**
Client Manager — **Sonia Uznadze**
Finished Artist — **Judy McLaughlan, Becca Soons**

DESCRIPTION

Ridley is the Australian based firm at the forefront of some of the biggest developments in the architecture and construction industry. Historically an architectural documentation specialist, they have pioneered the use of Virtual Design, using Building Image Modelling software (BIM).

With an overview of the entire building process from start to finish, Ridley acts as central hub for the architects, builders, and developers. In essence they have an understanding of the bigger picture. The identity plays on this idea, using individual elements, partial views, and cropping to create the whole. The use of human-centric data provides further understanding of the company, its people and projects.

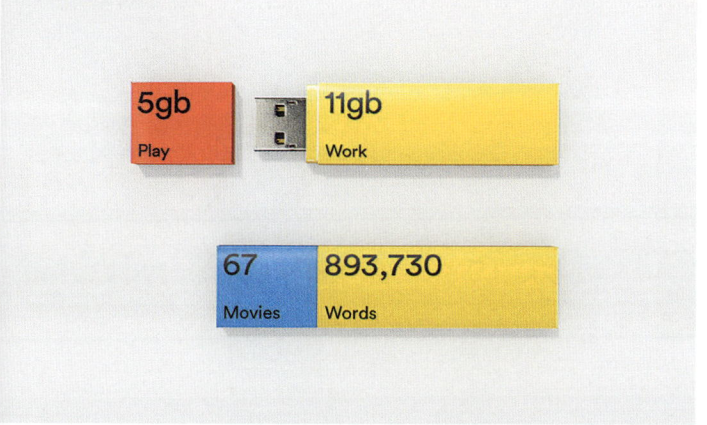

Client
exhibition
03

15 June 2014
Meeting room 3
12:00

Ridley

The VDCM Archives 2008—2015

BUILD ING INFORM ATION MODEL LING

THE USE OF BIM IN ENGINEERING PROJECTS

Ridley

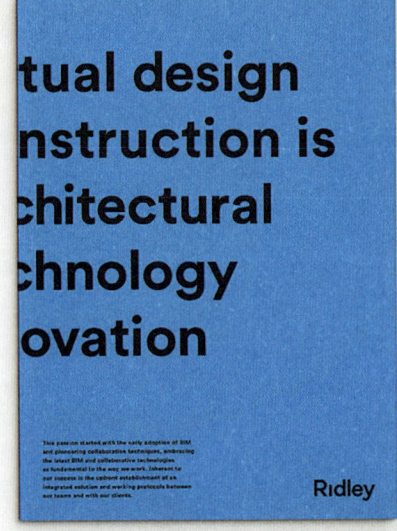

tual design nstruction is chitectural chnology ovation

Ridley

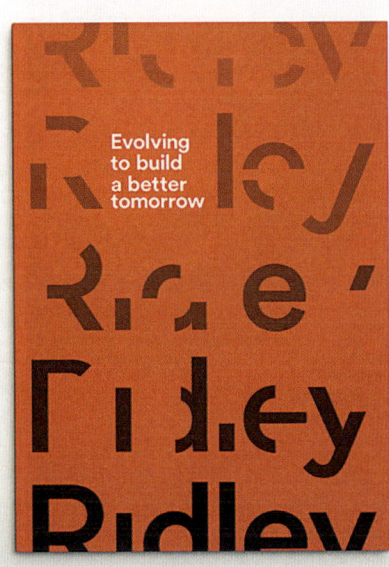

Evolving
to build
a better
tomorrow

Ridley

Sustainability
Report 2014

A growing global population places increased pressure on people and the environment. As the world develops, humanity places greater value on both life and the earth we all share. We believe that our social and environmental ecosystems are precious, and we're working to protect them.

Ridley

ae — Architecture and Engineering

vdcm — Virtual Design Construction Management

labs — Innovation Division

Ridley

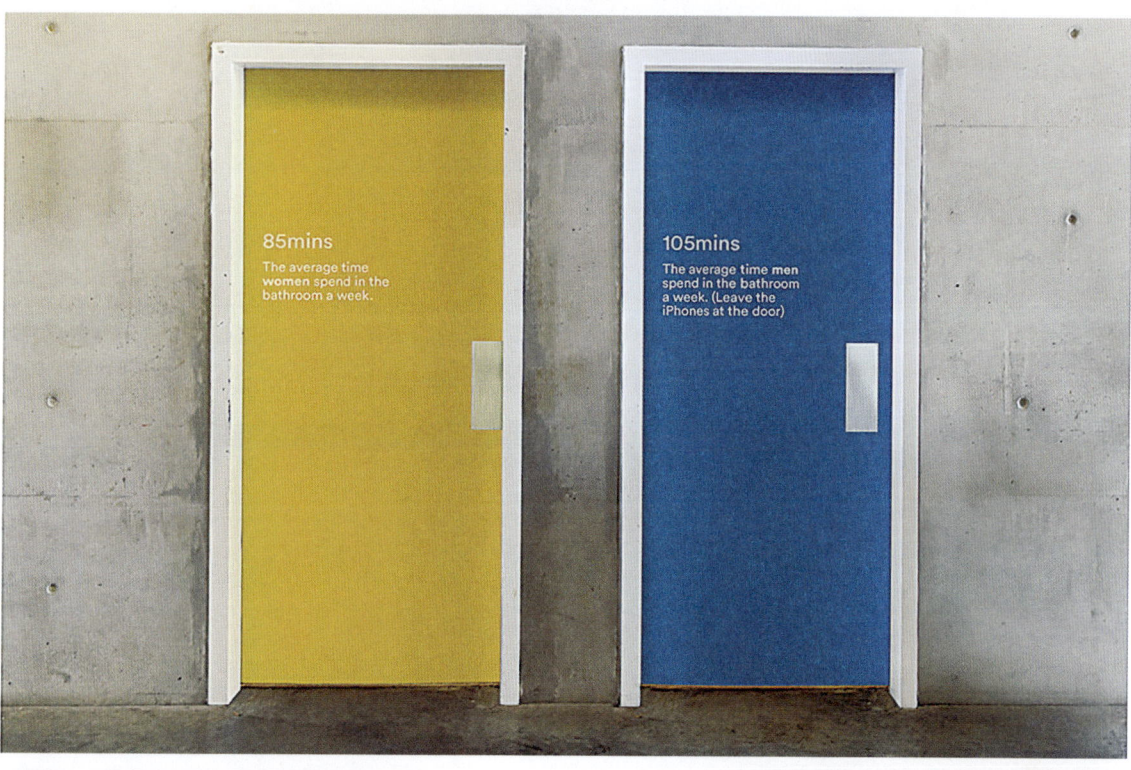

85mins
The average time **women** spend in the bathroom a week.

105mins
The average time **men** spend in the bathroom a week. (Leave the iPhones at the door)

URBANSCALE

TOWN & SOCIAL PLANNING
COLLABORATION
STRATEGY
SOLIDITY
DEVELOPMENT

2014
Shannon Meadows

URBANSCALE.COM.AU

URBANSCALE

Shannon Meadows, MPIA
Director
URBANSCALE

0423 016661
shannon@urbanscale.com.au
www.urbanscale.com.au

© URBANSCALE 2014.
Please do not reproduce without the
expressed written consent of URBANSCALE.

URBANSCALE

Urban
Planning
Consultants

Mobile: 0423 016 661
Mail: shannon@urbanscale.com.au
Website: www.urbanscale.com.au

Dolor sit amet, consectetur adipisicing elit, sed do eiusmod tempor incididunt ut labore et dolore magna aliqua. Ut enim ad minim veniam, quis nostrud exercitation ullamco laboris nisi ut aliquip ex ea commodo consequat. Duis aute irure dolor in reprehenderit in voluptate velit esse cillum dolore eu fugiat nulla pariatur. Excepteur sint occaecat cupidatat non proident, sunt in culpa qui officia deserunt mollit anim id est laborum.

Sed ut perspiciatis unde omnis iste natus error sit voluptatem accusantium doloremque laudantium, totam rem aperiam, eaque ipsa quae ab illo inventore veritatis et quasi architecto beatae vitae dicta sunt explicabo. Nemo enim ipsam voluptatem quia voluptas sit aspernatur aut odit aut fugit, sed quia consequuntur magni dolores eos qui ratione voluptatem sequi nesciunt. Neque porro quisquam est, qui dolorem ipsum quia dolor sit amet, consectetur, adipisci velit, sed quia non numquam eius modi tempora incidunt ut labore et dolore magnam aliquam quaerat voluptatem. Ut enim ad minima veniam, quis nostrum exercitationem ullam corporis suscipit laboriosam, nisi ut aliquid ex ea commodi consequatur? Quis autem vel eum iure reprehenderit qui in ea voluptate velit esse quam nihil molestiae consequatur, vel illum qui dolorem eum fugiat quo voluptas nulla pariatur.

But I must explain to you how all this mistaken idea of denouncing pleasure and praising pain was born and I will give you a complete account of the system, and expound the actual teachings of the great explorer of the truth, the master-builder of human happiness. No one rejects, dislikes, or avoids pleasure itself, because it is pleasure, but because those who do not know how to pursue pleasure rationally encounter consequences that are extremely painful. Nor again is there anyone who loves or pursues or desires to obtain pain of itself.

Kind regards,

SHANNON MEADOWS
Director

URBANSCALE

CREDITS

Studio — **makebardo**

DESCRIPTION

URBANSCALE is a planning consultancy based in Melbourne and Ballarat, Victoria. They specialise in planning strategies and planning systems, and they conduct research that explores urban dynamics and change. Their clients are local governments and consultant teams responding to complex urban planning and community challenges.

The interesting thing about working with a startup is that the work environment is flexible and close knit. Together with Shannon Meadows, Founder and Director of Urbanscale we worked from the beginning as a team to respond to their needs as new market entrants, stressing that though new, they have a solid base of experience, professionalism and efficiency. We are convinced that startups offer the opportunity to improve the lives of people in a tangible way and Urbanscale is one of them.

We worked with the idea of scale and coordination. The synthesis of measurement objects and scales were used as resources to be incorporated into the logo, intensifying the relationship between the words "urban" and "scale."

The "S" takes leadership to allow the game of differentiation between words and add conceptually the idea of unity, connection and coordination, making scale in typography.

As for the font choice, we worked on the redesign of a san serif typeface, to give the brand a unique character. For this reason we adapted details of scales and measures creating a new typeface to highlight the balance between the parties.

The colour palette used was a duo-tone of grey at 95% with vibrant yellow Pantone, resulting in a very interesting game because they are colours that coexist perfectly in the context where the brand will live. Yellow is a colour that gives soul, is vibrant and above all has the characteristic of being strong and full of character. The grey is the perfect companion to balance the scream of the yellow.

As a result of this first proposal, we have achieved the goal of creating a brand following every detail, aligning values such as cooperation, efficiency and professionalism.

Over all it is an honest logo, because with good typographic work we conveyed a solid and effective idea.

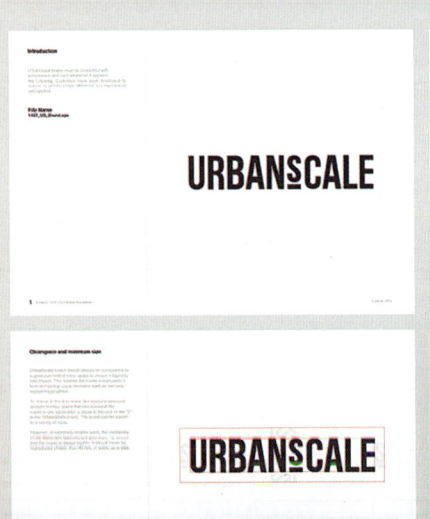
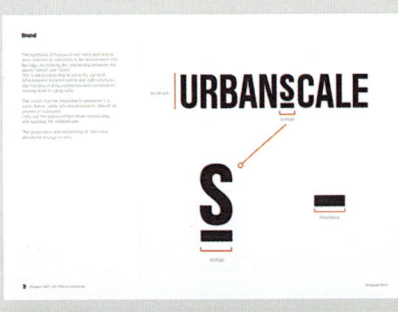

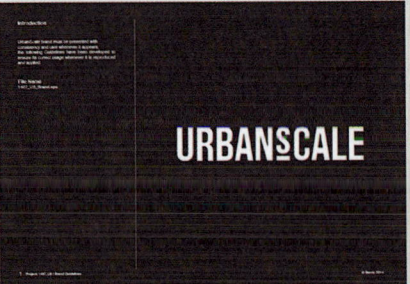

URBANSCALE

Home   About   Expertise   Case Studies   Let's Talk   #Planchat

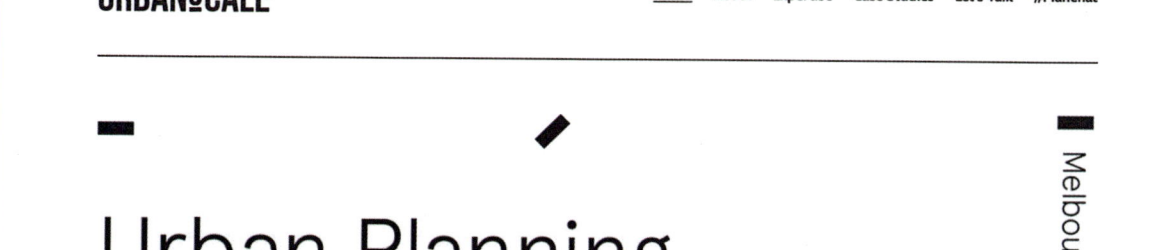

Urban Planning Consultants

Melbourne + Ballarat

URBANSCALE is a planning consultancy based in Melbourne and Ballarat, Victoria. We specialise in planning strategies and planning systems, and we conduct research that explores urban dynamics and change. Our clients are local governments and consultant teams responding to complex urban planning and community challenges.

VIEW MORE

Level 17, 31 Queens Street, Melbourne 3000
© **URBANSCALE** 2015

urbanscale.com.au

INTRA DESARROLLOS

CREDITS

Studio — **FIRMALT**
Creative Director — **Manuel Llaguno**

DESCRIPTION

Intra Desarrollos is a fully integrated real estate development company with more than 65 years of experience based in the city of Veracruz, Mexico. Since its inception, they have been responsible for the design, financing, and building of major developments in the industrial, hospitality, public and residential industries.

Intra Desarrollos works every day to continue innovating and adapting to new trends and technologies to continue to create developments that have positive impact in their respective communities.

When they approached us, they wanted to reinvent their brand and have it convey a message of strength and security for its investors and homeowners. The mark was designed to resemble a shield, a timeless emblem that has come to represent integrity and dependability.

When the shield is seen sideways, it displays the letters "I" and "D" of Intra Desarrollos. This came to be the perfect symbol to convey their core values of responsibility, sustainability and social responsibility.

OLIVER STEER INTERIOR DESIGN

CREDITS

Studio — **Phage Ltd.**
Designer & Creative Director — **Natasha Zlobec, Danny Brooks**

DESCRIPTION

Interior designer Oliver Steer commissioned Phage to redesign his brand to more accurately reflect the high calibre of his client-base, consolidating his visual identity, and differentiating him from local competition in the North West of England.

Point of sale and printed marketing materials were designed to reflect Oliver's own interior design style, with all the trappings of luxury necessary to appeal to his target audience. A luxurious brochure and presentation folder were produced, with accompanying design case study cards and company profile enabling items to be assembled and distributed in a range of combinations according to target audience, promoting clearer brand communication and better filtering of enquiries.

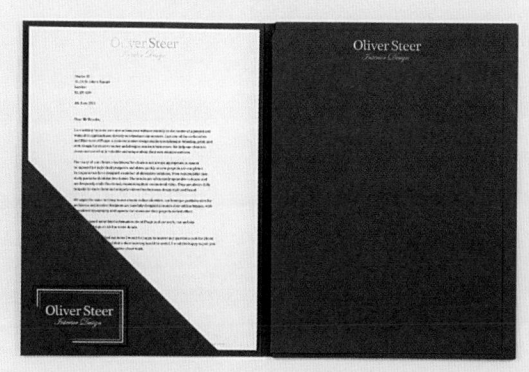

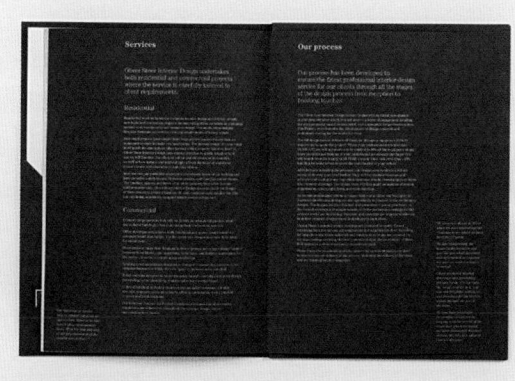

THOMAS MAE
Property & development

THE BENEFITS OF
A LAND & NEW
HOMES BUSINESS
Are you aware of the opportunities?

LAND & NEW HOMES

GENERATE LAND
FEE INCOME

GROW YOUR
BUSINESS INTO
NEW AREAS

INCOME AND PROFIT
FROM NEW HOME SALES

GROW YOUR
LETTINGS BUSINESS

THOMAS MAE

CREDITS

Studio — **Workbrands**
Account Manager — **Emma Jones**
Art Director — **Tom Ovens**
Creative Director — **Nick Farrar**
Designer — **Simon Tandy**
Copywriter — **Simeon de la Torre**

DESCRIPTION

With over 12 years of experience at Romans, Thomas Mae's founder Kevin Ellis has a wealth of insider knowledge. He commissioned Workbrands to create a strong brand that would help get his new consultancy business get off to a flying start.

A logo mark was created alongside a bespoke icon set representing the five key service offerings. This was then applied to stationery, company brochures, sales tools and marketing materials. It was important for Thomas Mae to be seen as contemporary, yet established, and to fully convey the breadth of services on offer.

Workbrands subsequently worked on the design and build of a brand new responsive website for Thomas Mae, which has helped reinforce their high-end brand perception and cement their position in the marketplace.

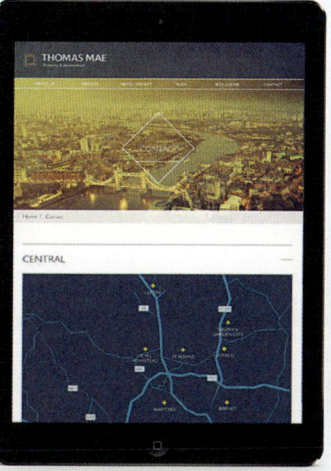

CHRISTOPHER ELLIOTT

CREDITS

Studio — **StudioBrave**
Creative Director — **Tim Sutherland**
Designer — **Carlo Mussett**

DESCRIPTION

Christopher Elliott approached us at time when his interior design profile was making waves in the industry as an emerging talent. With recognition growing, the time was right to establish a distinctive identity, both robust and restrained, reflective of his core principles.

We began with a brand workshop to define the unique attributes of the practice. A theme that emerged was that of "harmonious contrasts." While ironic in essence, they are characteristics which differentiate the practice's style. The meticulously crafted identity combined juxtapositions of subtle and bold elements, and harmony of nature and geometry.

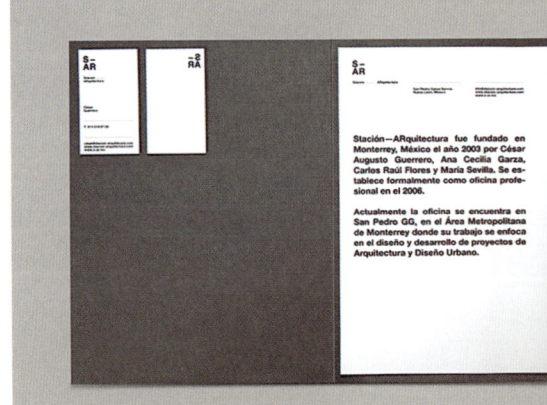

Stación—ARquitectura fue fundado en
Monterrey, México el año 2003 por César
Augusto Guerrero, Ana Cecilia Garza,
Carlos Raúl Flores y María Sevilla. Se es-
tablece formalmente como oficina profe-
sional en el 2006.

Actualmente la oficina se encuentra en
San Pedro GG, en el Área Metropolitana
de Monterrey donde su trabajo se enfoca
en el diseño y desarrollo de proyectos de
Arquitectura y Diseño Urbano.

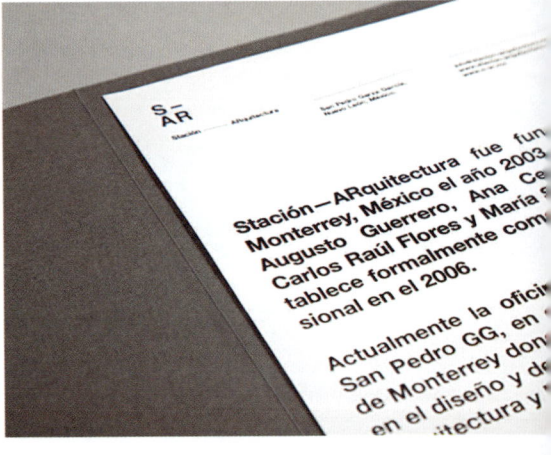

Stación—ARquitectura fue fun
Monterrey, México el año 2003
Augusto Guerrero, Ana Ce
Carlos Raúl Flores y María 5
tablece formalmente com
sional en el 2006.

Actualmente la ofici
San Pedro GG, en
de Monterrey don
en el diseño y d
itectura y

STACIÓN-ARQUITECTURA

CREDITS

Studio — **SAVVY STUDIO**
Creative Director — **Rafael Prieto**
Art Director — **Eduardo Hernández**
Designer — **Ricardo Ojeda**
Photographer — **Savvy Studio**

DESCRIPTION

Since 2003, Stación-ARquitectura has been working on innovative architecture and urban design projects in Mexico. The brand identity for this forward-thinking studio is based on their own design philosophy, which centres on creating clean and uncomplicated architectural environments that solve complex problems in simple ways, allowing the studio to concentrate their energy in generating functional, ample spaces.

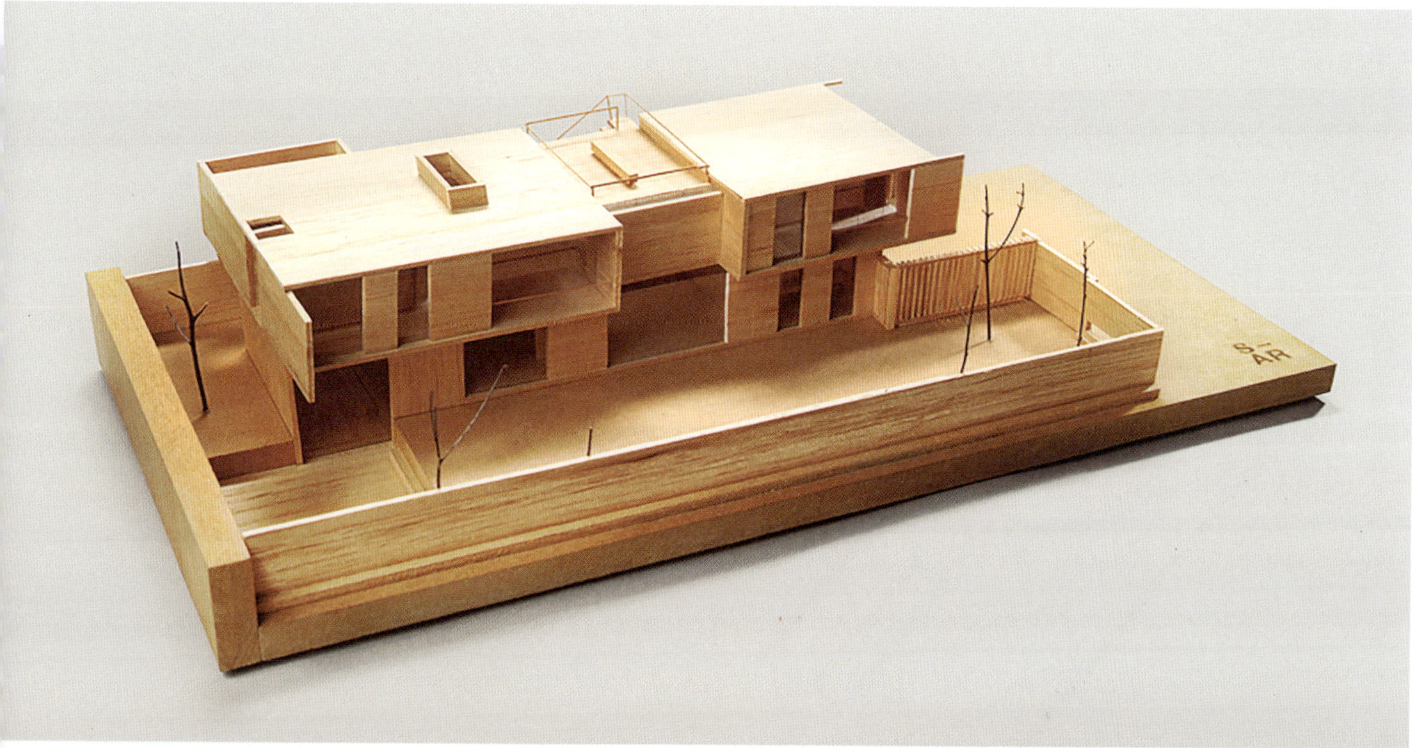

ULTIMATE TEAM

CREDITS

Studio — **Nhomada**
Project Director — **Diego Leyva**
Photographer — **Ana Georgina**
Graphic Designer — **Diego Leyva**

DESCRIPTION

Ultimate Team is a construction management firm
based Miami, Florida. The source for construction,
remodelling and maintenance issues of the area.
My approach was to create a contemporary,
elegant and timeless brand. Jose Torres
owner of the company wanted to stand out
his Mexican roots, that is why the logo itself
is a Mayan/Olmec face. The project includes
identity, naming, stationery and printing.

CM2

Studio — **The Negra**
Project Director — **The Negra**
Creative Director — **The Negra**

DESCRIPTION

We developed the complete branding for Cm^2, an architecture studio based in Buenos Aires and Santiago de Chile that designs minimal and modern environments where people want to be. The challenge was to develop a branding that expresses their values: Modernism, Simplicity, Good Design and Professionalism.

In order to achieve this, we decided to build a beautiful, minimal gridded layout based on the Dieter Rams concept: "Good design is as little design as possible."

The result? An identity based in less-is-more concept, black & white colour palette with emphasis on typography that reflects the studio spirit.

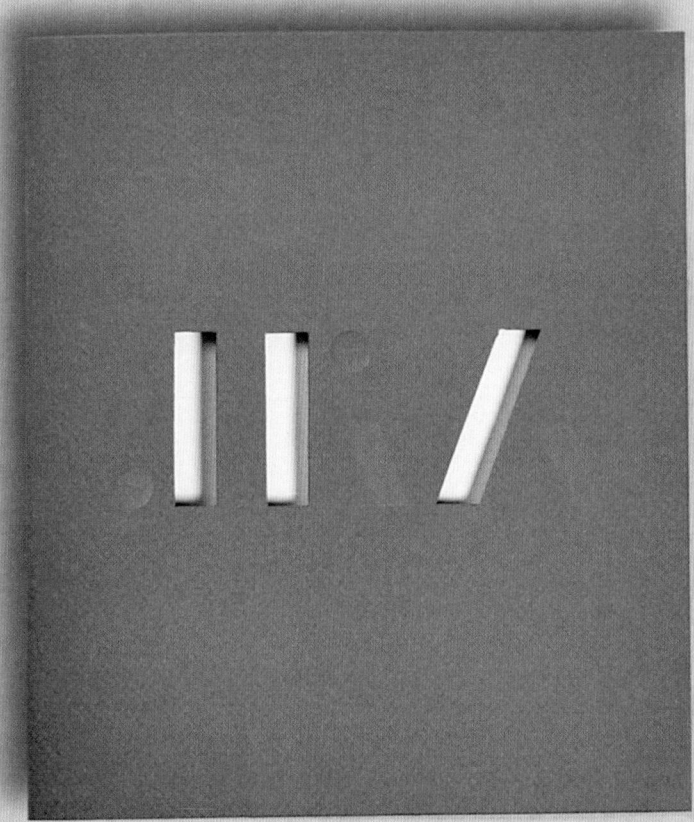

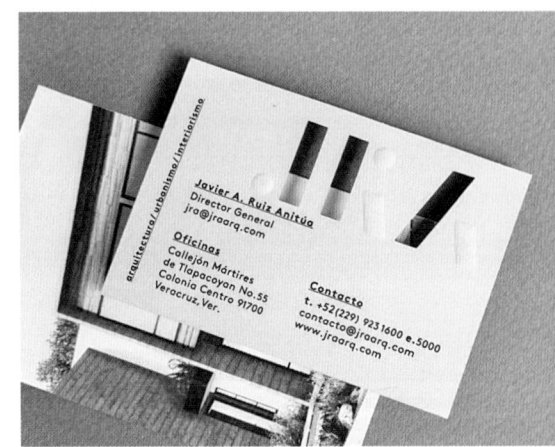

JRA

Studio — **FIRMALT**
Creative Director — **Manuel Llaguno**
Lead Designer — **Déborah G. Neaves**

DESCRIPTION

JRA is a professional and reliable firm specialized in architectural design located in Veracruz, Mexico. They are known for their exclusivity and vanguardism in each project. They work in different scales and sectors, providing solutions in the areas of integral architecture development, urbanism, interior, and landscape.

Our branding proposal takes inspiration in the contrast found in light and shadows. Die cuts and embossing are used throughout the brand's collateral, honouring architects' play with space and dimensions.

The logotype is an abstraction of the firm's name — JRA. Custom characters were designed to portray volume and different perspectives of the same space.

Shades of grey and white were chosen as JRA colour palette, again making reference to light and shadow. The sobriety of the colours helps the brand maintain a low-profile, with the idea that the aesthetic does not intervene with their projects.

The brand's pattern was developed as a complement to the elegance of JRA, embellishing the brand with texture.

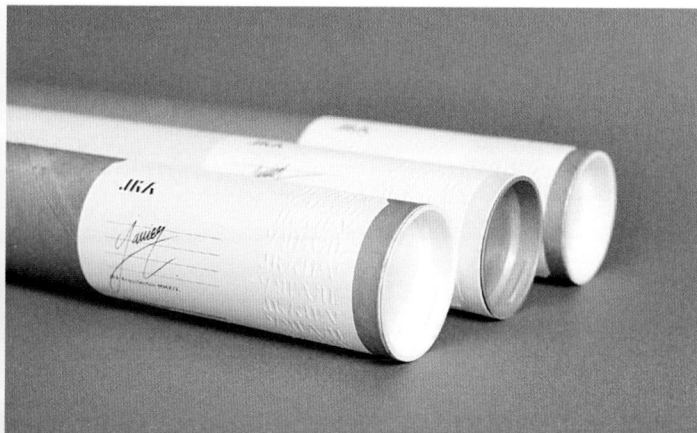

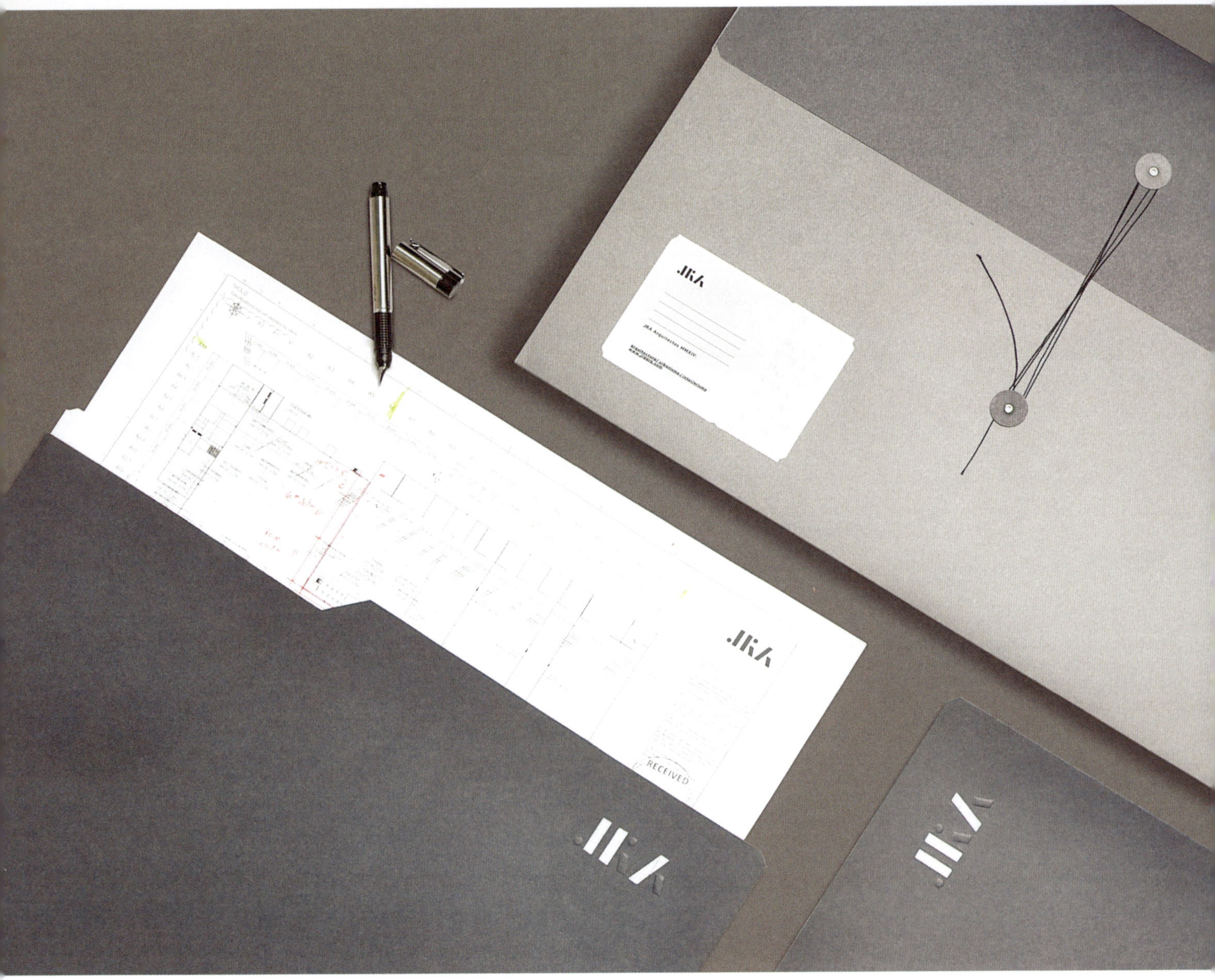

Javier A. Ruiz Anitúa
Director General
jra@jraarq.com

Contacto
t. +52(229) 923 1600 e.5000
contacto@jraarq.com
www.jraarq.com

Oficinas
Callejón Mártires
de Tlapacoyan No.55
Colonia Centro 91700
Veracruz, Ver.

arquitectura / urbanismo / interiorismo

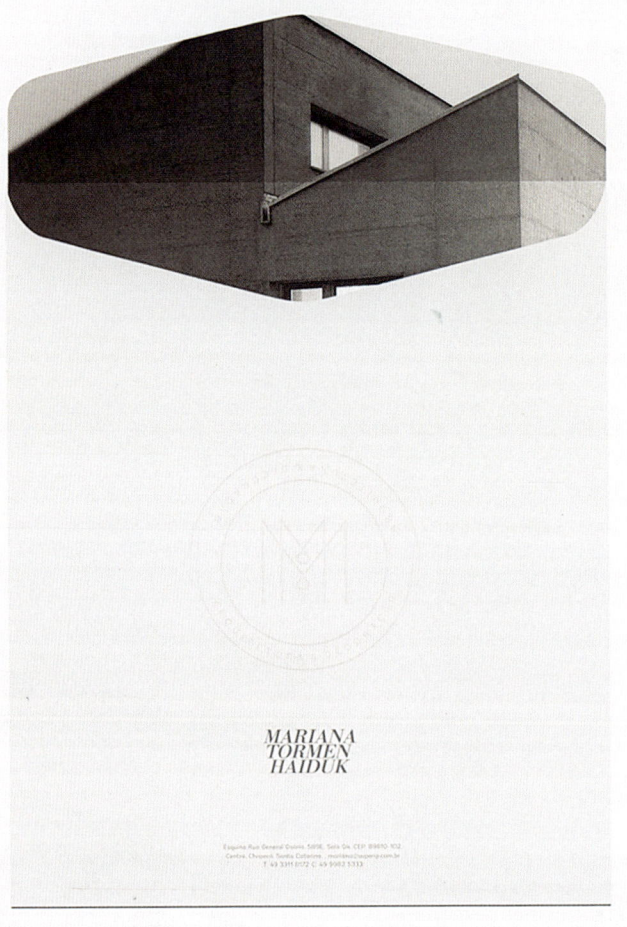

MARIANA
TORMEN
HAIDUK

MARIANA
TORMEN
HAIDUK

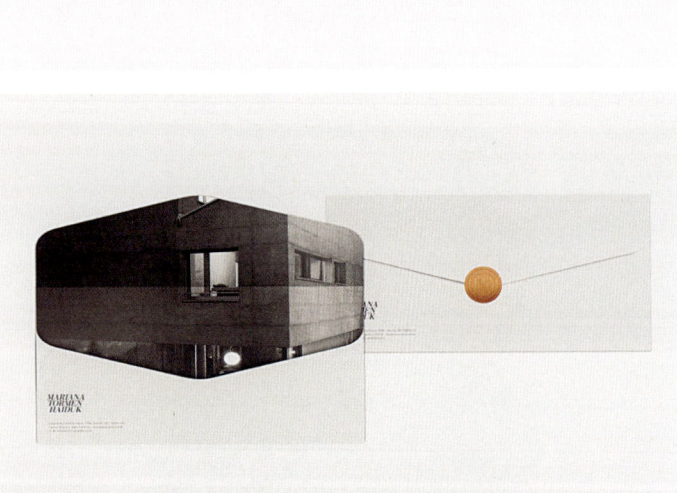

MARIANA TORMEN HAIDUK

CREDITS

Studio — **Estúdio Alice**

DESCRIPTION

Mariana Tormen Haiduk is an up-and-coming architect in the state of Santa Catarina. Up till recently, her career was based on collaborations with other architects and architecture firms. In 2012, she decided to open her own firm and hired Estúdio Alice to create her visual identity. Alice tried to identify if there were recurring elements in Haiduk's architecture projects.

This would be essential for her logo. Then we found out she has a lot of styles, so our challenge became to define this variation in graphical form. Even though we had lots of reference work — and therefore too much information — we achieved some design values by simplifying forms, choosing a very polished font and attributing some personality to the icon, based on Haiduk's initials. We wanted more than a logo — we wanted a signature.

We went for high-contrast colours, with a touch of handicraft, organic and, when necessary, customizable elements. It is Haiduk's way of saying that each architecture project should be singular.

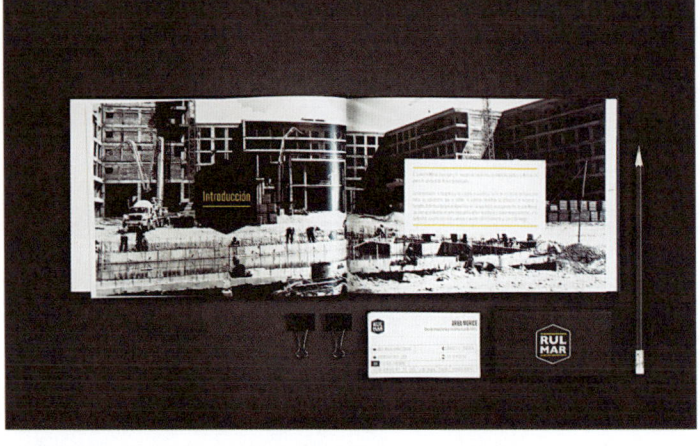

RULMAR

CREDITS

Studio — **Menta Picante**
Project Director — **Gabriela Salazar**
Creative Director — **Alejandro Román**

DESCRIPTION

Project that involves graphic identity, website and brochure for a construction company located in Panamá.

Rulmar is a Mexican company, with headquarters in Panamá, dedicated to the planning and construction of projects, focusing on their knowledge and efforts to solve the needs of their costumers with high quality solutions in less time. Their main customer is the hotel chain RIU.

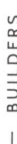

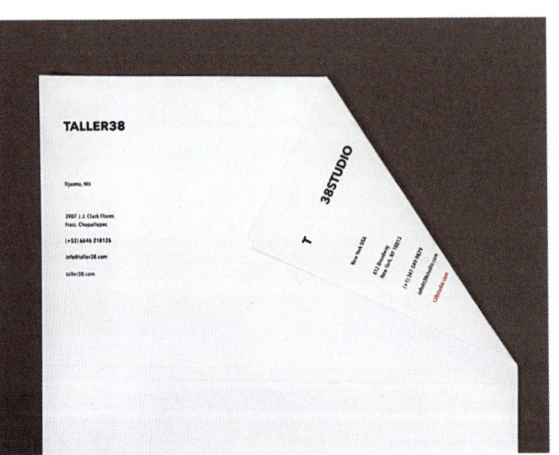

TALLER38

Studio — **SAVVY STUDIO**
Creative Director — **Raúl Salazar**
Account Director — **Orlando Fernández**
Art Director — **Eduardo Hernández**
Designer — **Bernardo Domínguez**

DESCRIPTION

Taller38 (T38) is an office dedicated to developing real estate projects. T38 is a multidisciplinary workshop composed not only by architects, but also by artisans, carpenters, painters and welders.

Together they develop architectural pieces and real estate projects that are a benchmark of high quality, with the aim to uplift the style and quality of life of the people who live there.

For both workshops, we worked with a creative concept based on modularity, always with a ruler based grid.

As for the identity, all the applications — from the logo to the web page — were designed and developed from a unit that is divided into two parts. On one han,d we created a unique identity for each of the two companies, but at the same time, it reflects the synergy between two entities that are part of a whole.

That is, if these applications were superimposed, all the elements would complement each other. To differentiate each other, every workshop was assigned with a complementary colour: red and blue, basic but opposite colours.

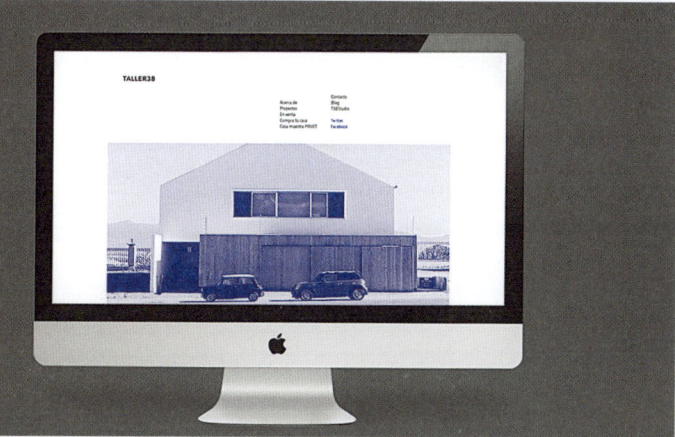

ARBENIGOL

Studio — **Actual Studio**
Graphic Designer — **Jonathan Ford**

Cardiff-based property development
consultancy Arbenigol provide advice
on all aspects of the development process;
from purchasing a site to building on it and
then selling it. Actual Studio were approached
to designed an identity for the new business,
as well as a website and accompanying
branded materials.

benigol
development

erty development

word for specialised. The
usiness' unique and bespoke
evelopment, whilst referencing
eritage

he property development
nore complicated due to added
benigol specialises in a pro
service, founded on honesty

ces we are able to guide

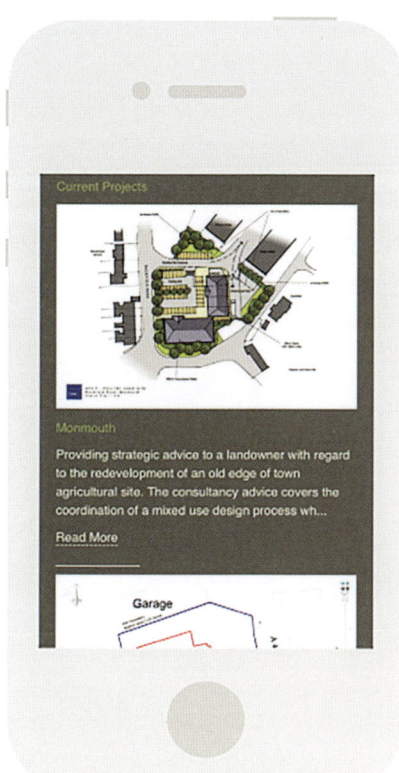

Current Projects

Monmouth

Providing strategic advice to a landowner with regard
to the redevelopment of an old edge of town
agricultural site. The consultancy advice covers the
coordination of a mixed use design process wh...

Read More

Garage

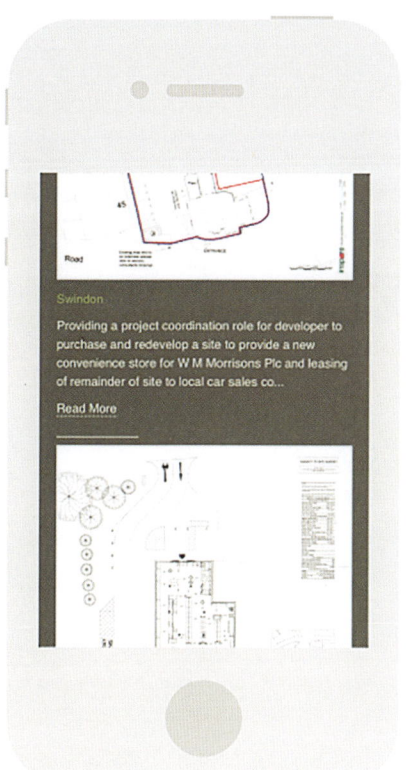

Road

Swindon

Providing a project coordination role for developer to
purchase and redevelop a site to provide a new
convenience store for W M Morrisons Plc and leasing
of remainder of site to local car sales co...

Read More

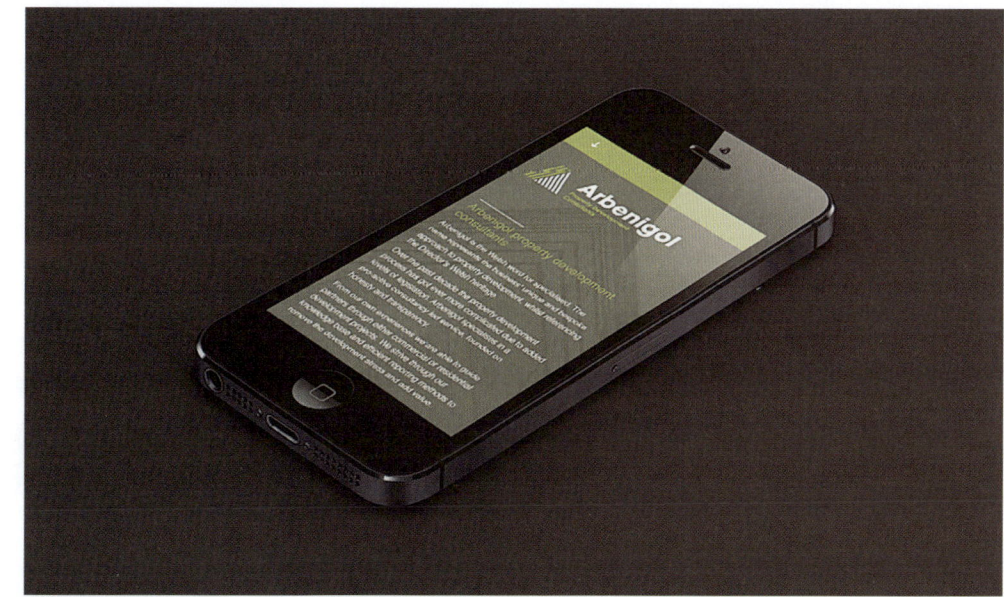

BENNETT & FOSKETT

CREDITS

Studio — **Brogue Studio**
Creative Director — **Nicholas Hards**
Creative Director — **Jonathan Hards**

DESCRIPTION

Bennett & Foskett are a modern property
renovation specialist with traditional values
and a passion for quality craftsmanship.
A new brand helped them build more credibility
and launch themselves into the market.

The dynamic duo who were in need of something
a little special to position themselves in the ever
demanding market, with one objective; modern yet
traditional, to reflect their own way of working,
their individual beliefs and the companies ethos.

Brogue Studio provided Bennett & Foskett
everything they needed to move forward in
their chosen market sector, from stationary
and custom vehicle signage to art direction,
promotional material and a responsive website.

B&F

+44(0)7725 348 082 / +44(0)7887 683 949
info@bennettandfoskett.co.uk
www.bennettandfoskett.co.uk
Property Renovation Specialists

Bennett&foskett

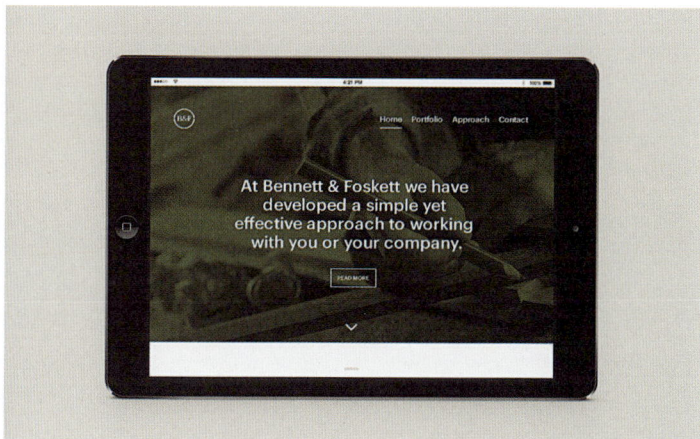

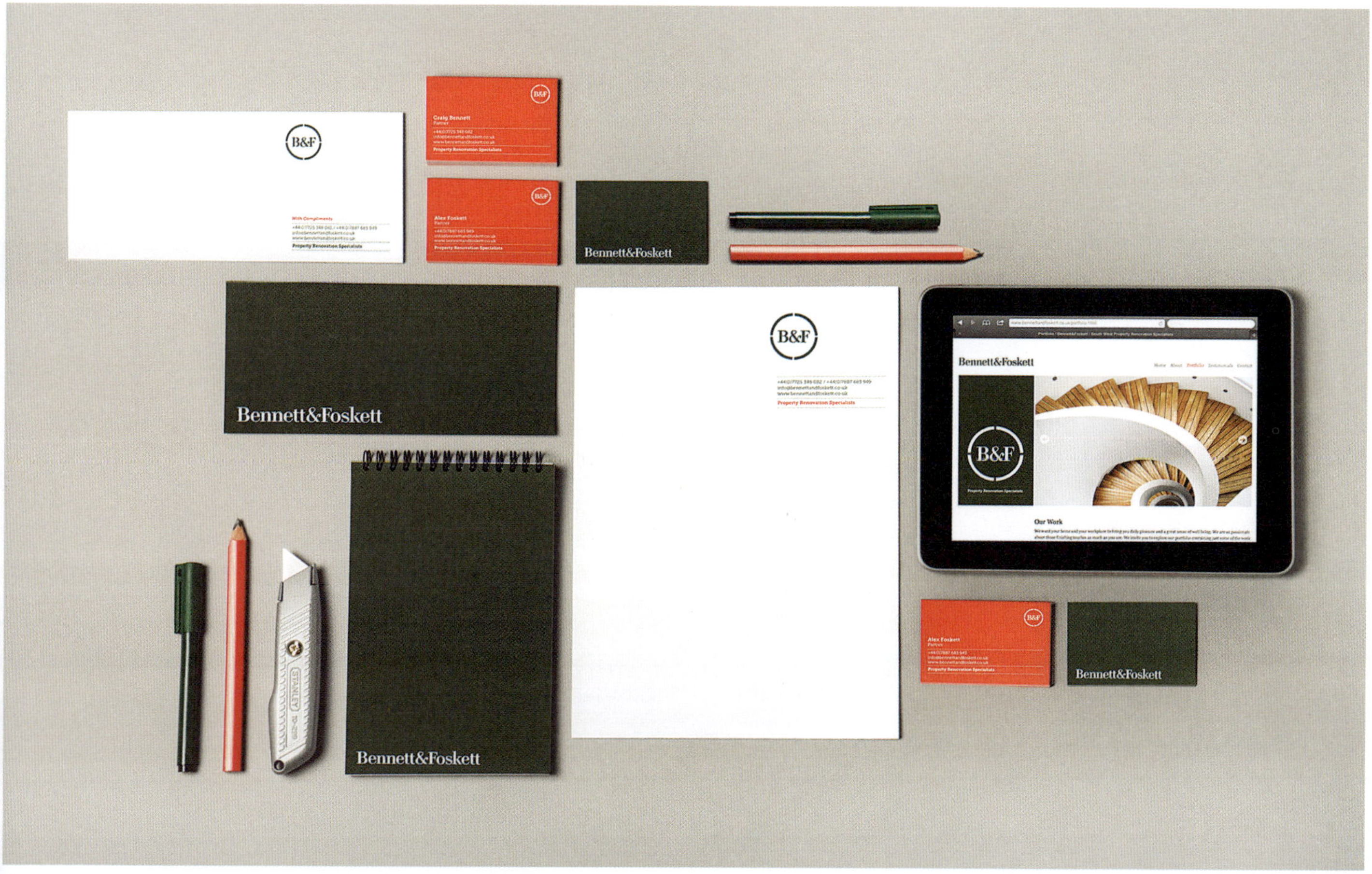

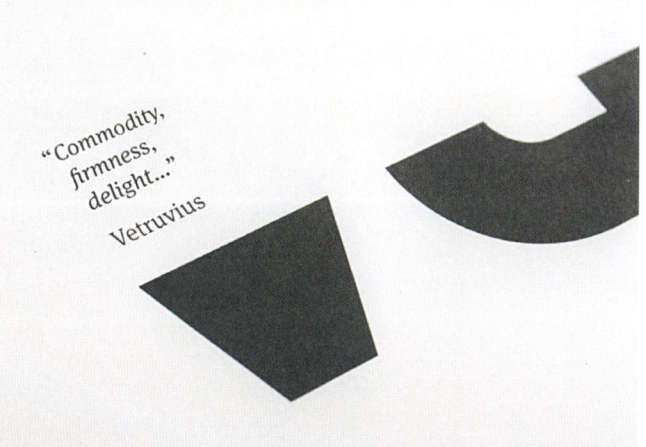

"Commodity, firmness, delight..." Vetruvius

PAPA

Studio — **Phage Ltd.**
Designer & Creative Director — **Natasha Zlobec, Danny Brooks**

DESCRIPTION

Award-winning architectural practice Papa came to Phage with a brief to design their new brochure and website; We took a broader look at their marketing as a whole, their work, and their target audiences.

The confident new identity we designed for them combined structural block type with an elegant modern serif, and used a range of materials and print finishes to communicate Papa's design-led approach and high quality built work to potential and existing clients. The printed practice profile is easily customisable in-house for different audiences, featuring a stunning high-gloss wrap and individual case study cards.

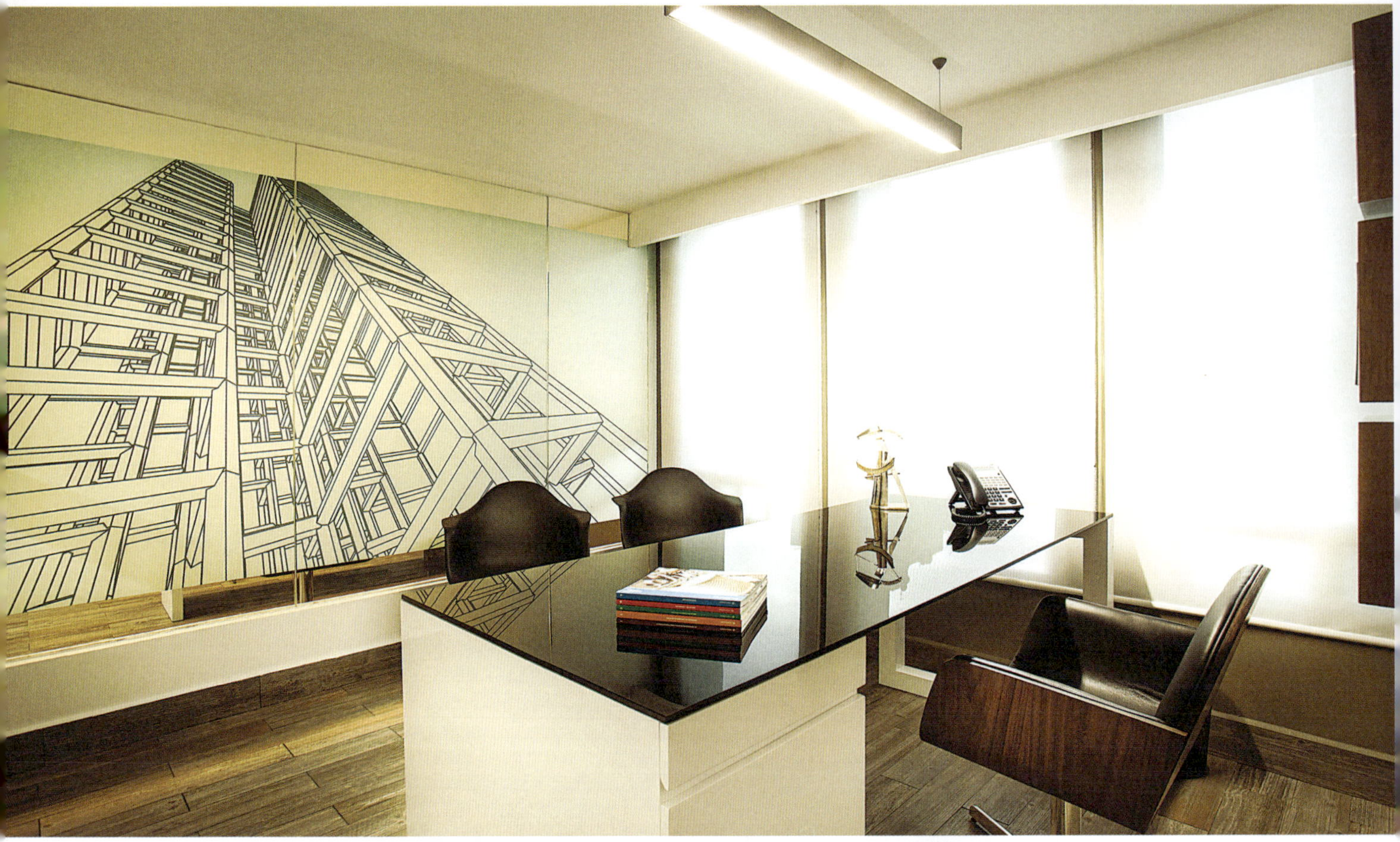

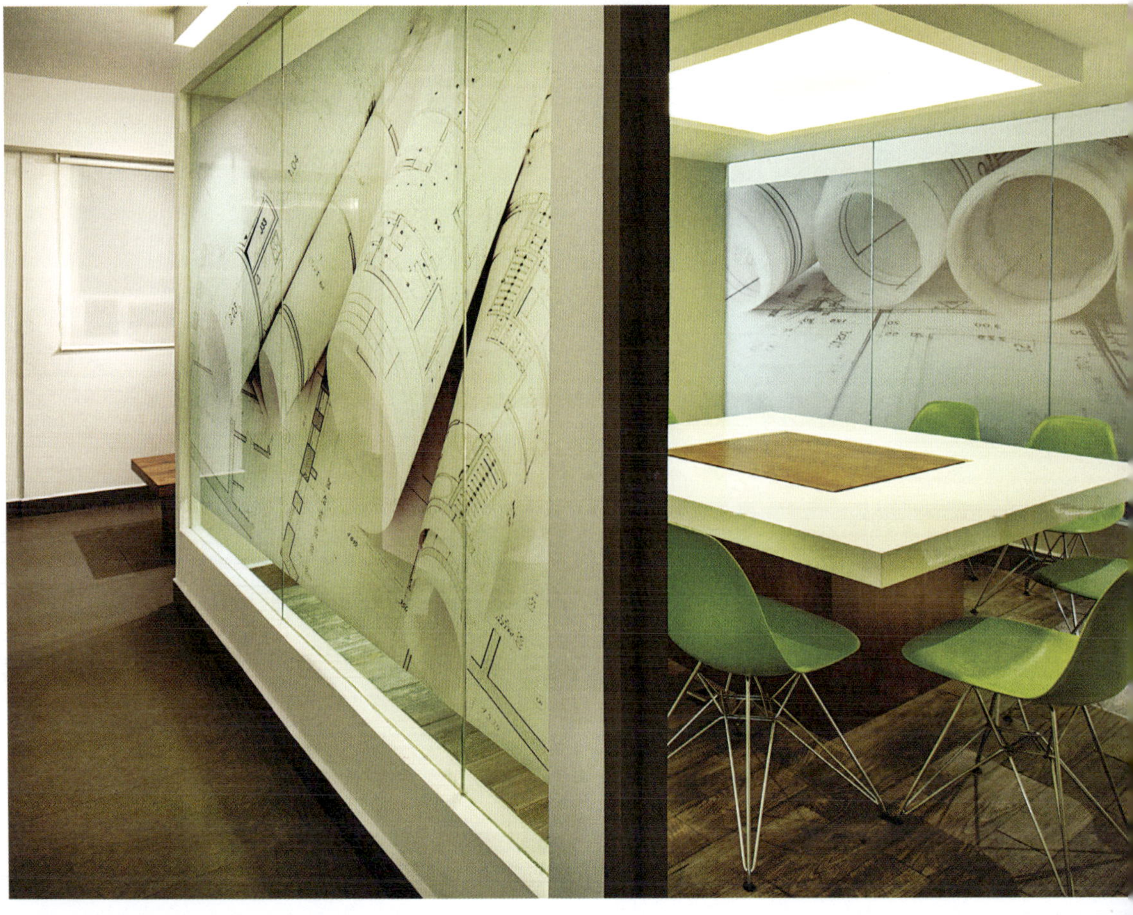

GRUPO BHAU

CREDITS

Studio — **Nhomada**
Creative Director — **Diego Leyva**
Photographer — **Ana Georgina**
Graphic Designer — **Diego Leyva**

DESCRIPTION

Grupo Bhau is a construction company based in Mexico City. Architecture, interior and furniture design are amongst the disciplines of this great Mexican brand. Branding and interface design were created for this client.

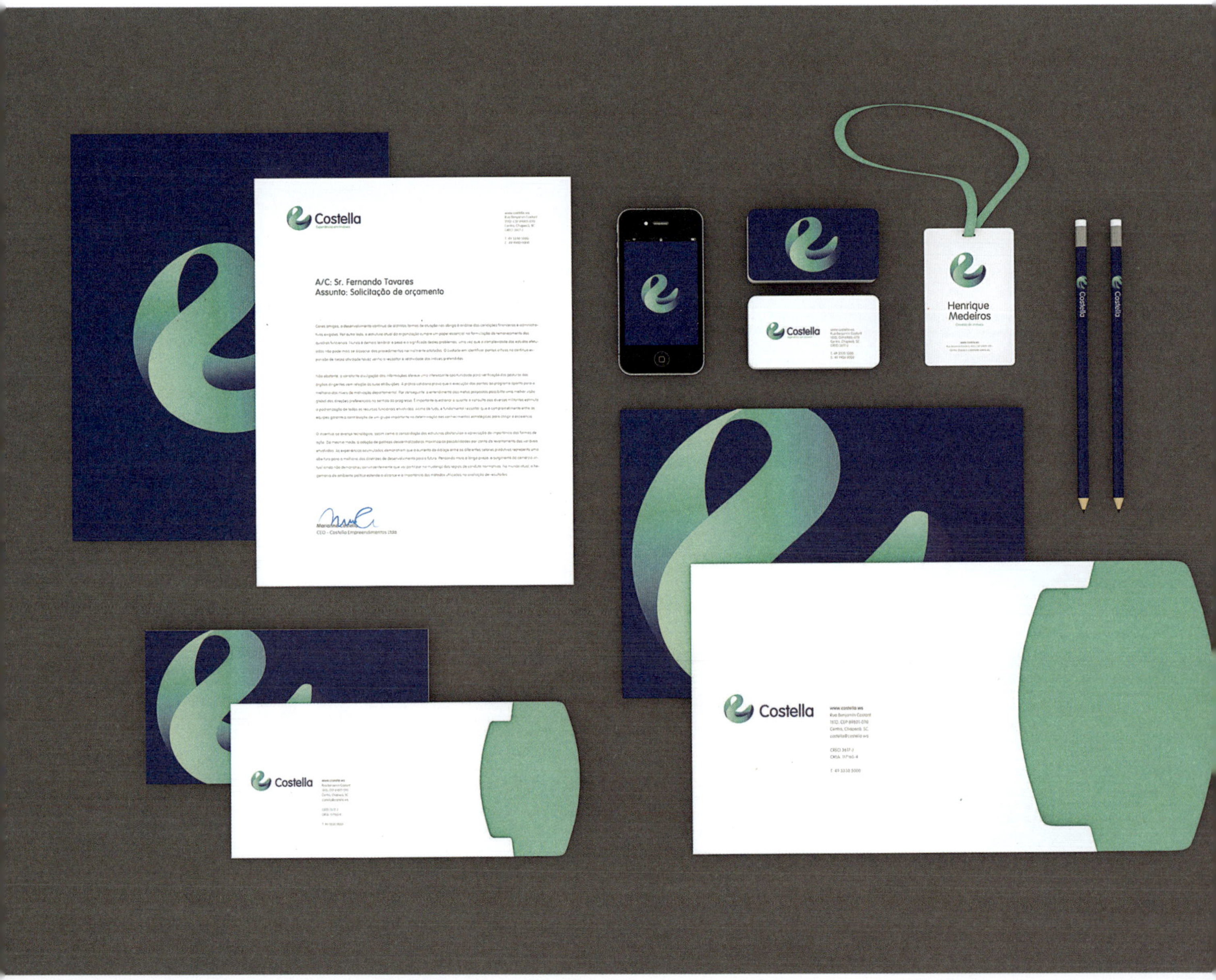

COSTELLA EMPREEN-DIMENTOS

CREDITS

Studio — **Estúdio Alice**

DESCRIPTION

Despite its origins in a family business that has been part of Chapecó's real estate development market for the last 30 years, Costella Empreendimentos was established as a brand new company. It is the same market segment the owners always dealt with, but the new company should be bolder and contemporary.

Estúdio Alice started working on this project when the new company was still in embryonic stage. Business-wise, everything was still a sketch. It was both challenging and perfect, since we were able to develop a new brand while a brand new company was being developed. We could step in and talk about positioning in a broad sense. That helped us define the company's stance in graphic form, as well as in other communication strategies.

We were all clear about what the company meant when we got to logo development. Costella should be a new option within a market segment that is dealing with new technology, evolving products and new safety requirements. But, mostly, it should deal with the well-being of families who want to live harmonically and sustainably with the environment. It is a positioning that turns the company into a trail-blazer for the local market. The logo should follow organic trends, getting away from all the squares and straight lines that are common in this segment. It should also be lighter — not only light colours, but also light shapes and light in its applications. These elements were decisive for the whole project.

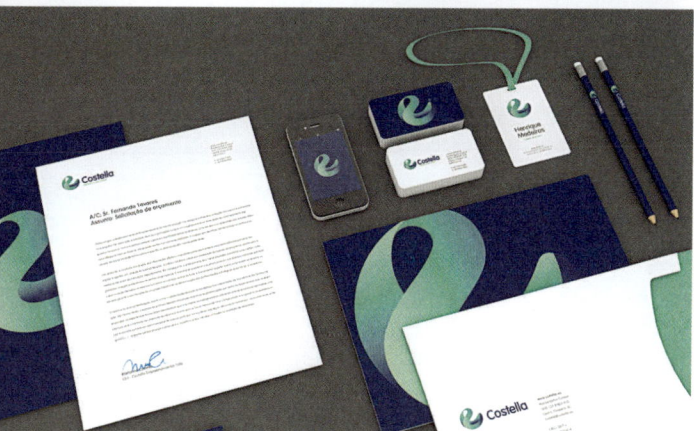

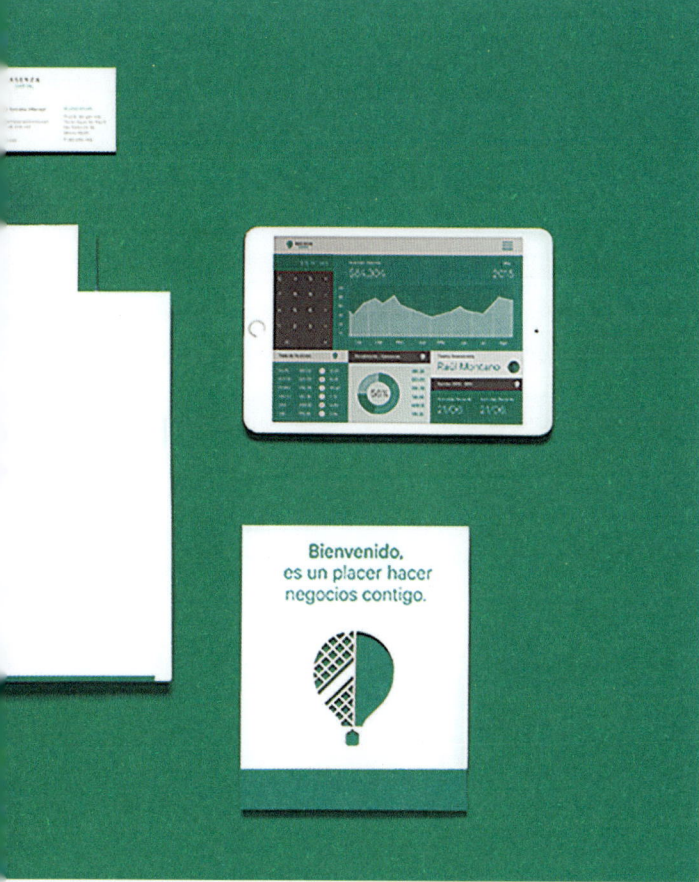

ASENZA CAPITAL

CREDITS

Studio — **FIRMALT**
Creative Director — **Manuel Llaguno**

DESCRIPTION

Asenza Capital is a venture capitalist based in Monterrey, Mexico specialized in real estate investments, public market, and private equity. Together with their associate developers, they seek to add financial and legal value to each project. Asenza provides solid strategies and networks for growth and development.

Elevation is what inspired us for the concept, the idea of taking businesses to a higher level. We combined elements of a shield and a hot-air balloon to design a mark that depicts security and altitude. The brand mark is strategically placed in elevated positions within the collateral, mirroring the ascent.

The brand is all about building trust without intimidating — Asenza wants to see their associates grow.

Branding elements include a geometric pattern inspired in the brand's mark. It gives the aesthetic an official feeling, and works harmoniously throughout all communications.

The typography intends to move away from your ordinary corporation, and offer a warm welcome to all. Green and gold compose the colour palette, as a representation of currencies and value.

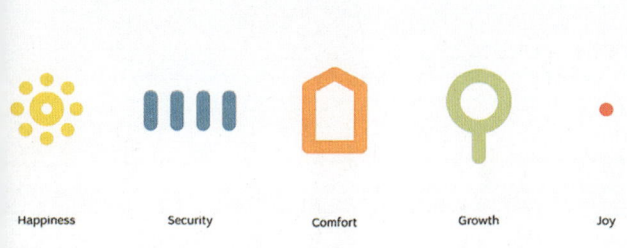

Happiness   Security   Comfort   Growth   Joy

HOME TO GO

CREDITS

Studio — **Hooga Creative**
Creative Director — **Dmytro Yarynych**
Art Director — **Denys Kuzmenko**
Designers — **Kyrylol Shvedov, Tetiana Kobryn**

DESCRIPTION

Getting yourself a place to live happily is now like ordering yourself a cup of coffee to go. Well, almost. home(to)go offers easy, quick and affordable solutions for housing, regardless of whether you have a piece of land. home(to)go promises to build your house for up to 40 days (and yes, you can move in right after that), using brand technology and best local materials and workforce. There are few basic designs available as well as custom projects.

Visual concept is as simple and accessible as the name — we tried to make the decision of building a new house comfortable, friendly and packed with positive emotions. On the other hand, we minimized the amount of printed material to keep the pricing as low as possible. We also made our best to incorporate the concept into the logotype itself.

CINEMA/THEATRE · RESTAURANT · SHOPPING · HORSEBACK RIDING · BIKE PATH · MUSIC VENUE · STONES/PAVERS · FURNITURE

COCKTAILS/BAR · LIBRARY/BOOK STORE · WATER ACCESS · GAZEBO · DOG PARK · GOLF COURSE · PALMS · GARDEN

MARKET · PLAYGROUND · PIEDMONT PARK · SIDEWALK/TRAIL · SCHOOL · COMPASS · FIRE PIT · ART GALLERY

TREES/NATURE · COMMUNITY POOL · DOWNTOWN · YMCA/TENNIS CLUB · CHURCH · BALL FIELD · FOUNTAIN · HOSPITAL

![Monte Hewett Homes logo]

JONES

CREDITS

Studio — **Actual Studio**
Graphic Designer — **Jonathan Ford**

DESCRIPTION

South-Wales construction firm Jones (formerly Jones Brothers) is a multi-discipline contractor working in areas such as Civil Engineering, Building Construction and Mechanical Engineering. We were approached to work with them on their re-branding.

CINEMA/THEATRE

RESTAURANT

SHOPPING

HORSEBACK RIDING

BIKE PATH

MUSIC VENUE

STONES/PAVERS

FURNITURE

COCKTAILS/BAR

LIBRARY/BOOK STORE

WATER ACCESS

GAZEBO

DOG PARK

GOLF COURSE

PALMS

GARDEN

MARKET

PLAYGROUND

PIEDMONT PARK

SIDEWALK/TRAIL

SCHOOL

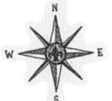
COMPASS

FIRE PIT

ART GALLERY

TREES/NATURE

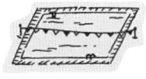
COMMUNITY POOL

DOWNTOWN

YMCA/TENNIS CLUB

CHURCH

BALL FIELD

FOUNTAIN

HOSPITAL

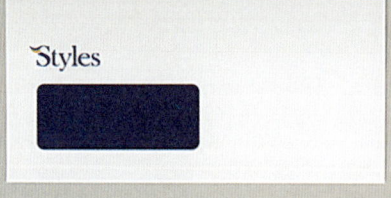
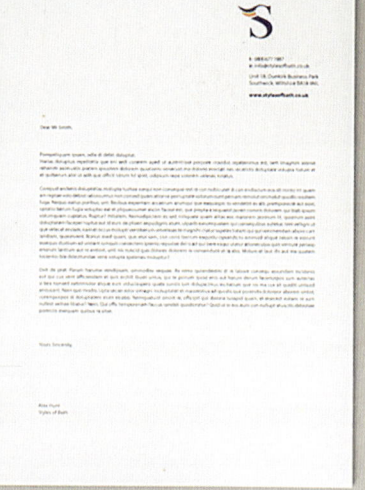

STYLES OF BATH

CREDITS

Studio — **Brogue Studio**
Creative Director — **Nicholas Hards**
Creative Director — **Jonathan Hards**

DESCRIPTION

Specialists in bespoke home solutions; Styles of Bath offer a wide range of services from property renovation through to bespoke windows and the design and implementation of unique outdoor spaces.

The implementation of the bold colour palette accompanied with a traditional serif typeface not only helped them to stand out, but cemented their creditability within an already crowded market sector.

The identity was carried across a wide range of mediums, from business cards and letterheads to work wear and direct marketing.

BUILDERS

LEVANTE

CREDITS

Studio — **Menta Picante**
Creative Director — **Alejandro Román**
Project Director — **Gabriela Salazar**

DESCRIPTION

Levante is a new real estate company located in Guadalajara, their main values to communicate are: vanguard, and the warm and easy approach with the clients. We worked on an icon inspired by cardinal points, using this concept as a reference since lands and estates are located by its coordinates — a well known concept where they can relate. The typographical selection makes a great combination with the icon by being a minimalist and sans serif typography, where it does not steal the attention from the icon. Both elements represent the luxury and vanguard of the company without losing the warm approach with the clients.

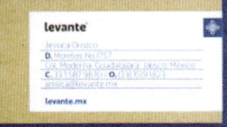

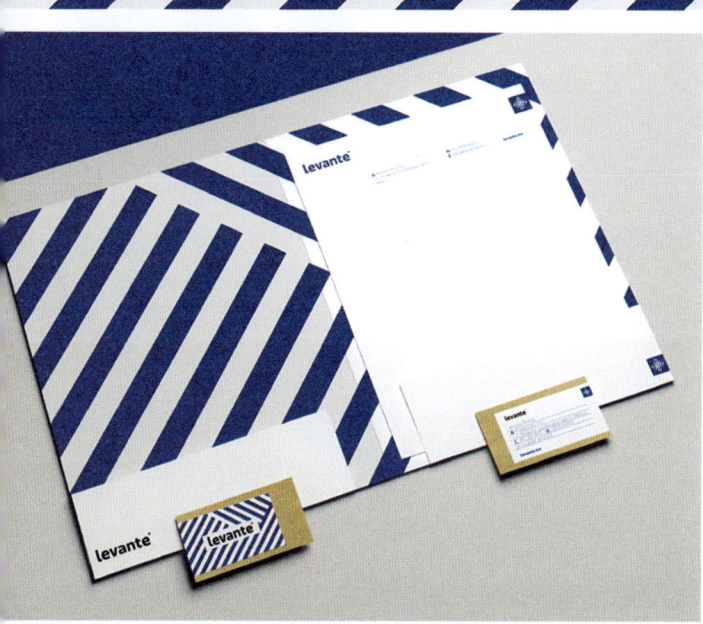

Future takes
the form

Building Exhibition'15
Krasnoyarsk

monolit-holding.ru

MONOLIT-HOLDING

MonolitHolding®
future takes the form

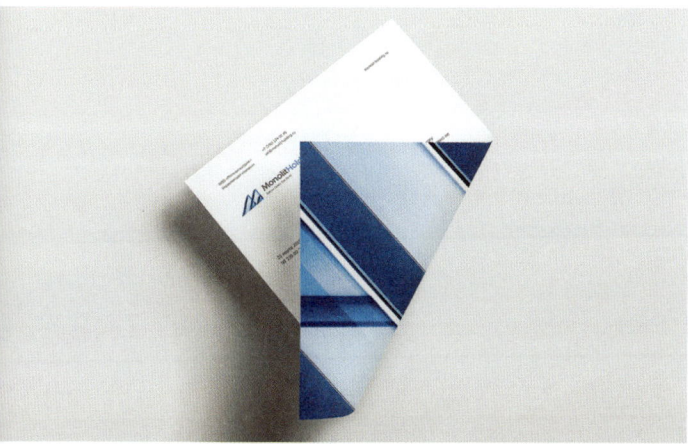

CREDITS

Studio — **SmartHeart® Agency**
Client — **Monolitholding**
Creative Director — **Stas Okruh**
Art Director — **Yuriy Mihalchenko**
Brand Strategist — **Anton Chitana**
Designers — **Ilya Tumaykin, Yuriy Mihalchenko**
Copywriters — **Anton Chitana, Danila Yusma ("Tverdiy znak")**
3D Visualisers — **Klyesov Sergey, Olga Tyryshkina, Elena Okruh**

Video (Logotype Animation)
Production Team — **Roma Lubimov, Dmitri Baraulin ("White Russian")**
Music — **"Radar"**

Brand Film
Production Team — **SmartHeart® Agency**
Music — **Long Arm — Blue Birds Red Flowers**

Brand Launch
Event Manager — **SmartHeart®, "Pryamaya liniya", "Tverdiy znak"**

DESCRIPTION

Creating new brand positioning for one of the leaders in the regional construction industry. The entire process took SmartHeart eight montsh of work and application of unique methods.

Stanislav Okruh, creative director of SmartHeart: "Not only friendship for ages but also common values connect us with 'Monolitholding.' This client is brave in his plans, he seek for innovations: the company wants to say its word in architectural design, it actively cooperates with Japanese and Swedish bureau. Recently 'Monolitholding' celebrated its 25th birthday. Management felt change in market conditions and new challenges, it decided to use re-branding as a point of grow for next 25 years. That is why we worked closely with the owners and 'the tops' on only possible level of management will, that requires revolution changes in companies."

Yuriy Mihalchenko, art director of SmartHeart: "When we work with branding of construction companies where time of active marketing brand life is 3–5 years we can't dive deep in the essence of developer. Re-branding of construction company is fundamentally different story, it requires deep intrusion. We had to get into heads of employees, to understand how look their ambitions, what inspires and motivates them. We analysed all competitors and correlated them with positioning of the client. We removed all the inconsistencies in strategic vision that was in the company and we got simple proportion: 'Monolitholding' is innovator for 70% and leader for 30%."

Modern image of builder of the future — leader and innovator — appeared as the result of several workshop with directors of the company. The main task of these brainstorms was scarch identity of the company.

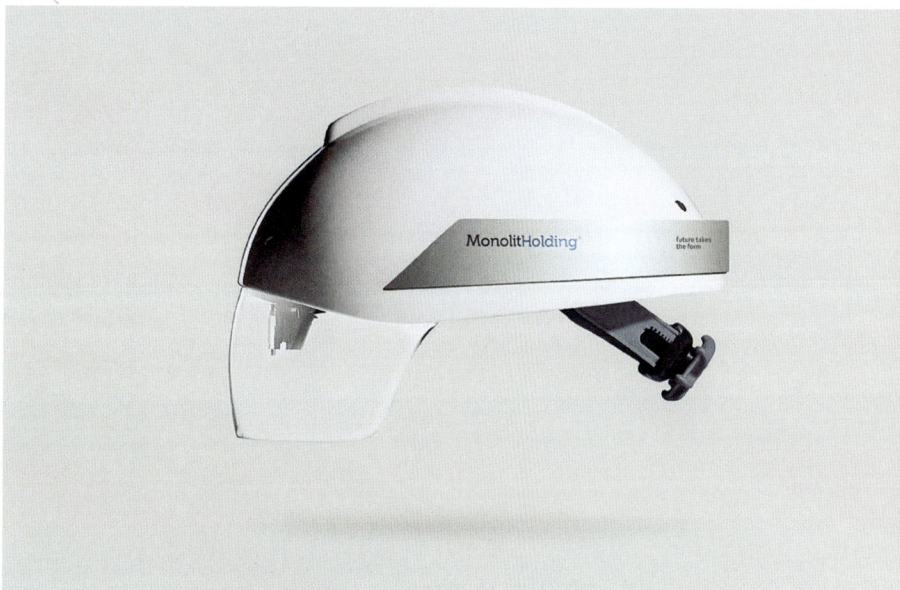

Kazim Abasov, Deputy Chairman
of the Board Monolitholding:
"It is important that aims of our company are
clear and values are transparent to people.
That is why new logo looks like glass. Keeping
in touch with old logo it reflects openness of
'Monolitholding,' it is brave and it is focus on the
future. This logo, fonts and colours embody all
that we love in our work. Blue colour is energy
of leadership and reflection of sky in façades of
buildings. Grey is colour of metal, embodiment of
innovations and high technologies. Glass is symbol
of transparency and openness to the world."

In the project brand book of the company,
it states the positioning, philosophy of the
brand and its main values. And it also contains
the rules about using of brand identity.

SmartHeart created a new innovative logo
and a recognizable communicative language
based on approved positioning by client. The
new brand communicate with the audience in
terms of innovation technologies and openness.
Communicating with consumers, there are
visual motifs like glass façades, friendly
faces looking through window and etc.

Visual language supported by new verbal
meanings. The company had a new slogan "Future
takes the form," a new templates of business
communication, big presentation booklet (which
was made in collaboration with copy-writing agency
"Tverdiy znak") with the possibility of translating
to foreign languages as "Monolitholding" works
with the Japanese and Swedish, where a part of
the promotional messages intrigues customers by
hieroglyphs and Scandinavian diacritical symbols.

All directions and holding project by principles
of innovations and leadership. SmartHeart hold
partial repositioning and found new focus in
construction objects. As a result, synchronization
of brand architecture appeared, in which
each object of "Monolitholding" is the most
advanced at the point of view of progress.

future takes
the form

A new neighborhood
"Preobrazhensky"

MonolitHolding®
future takes the form

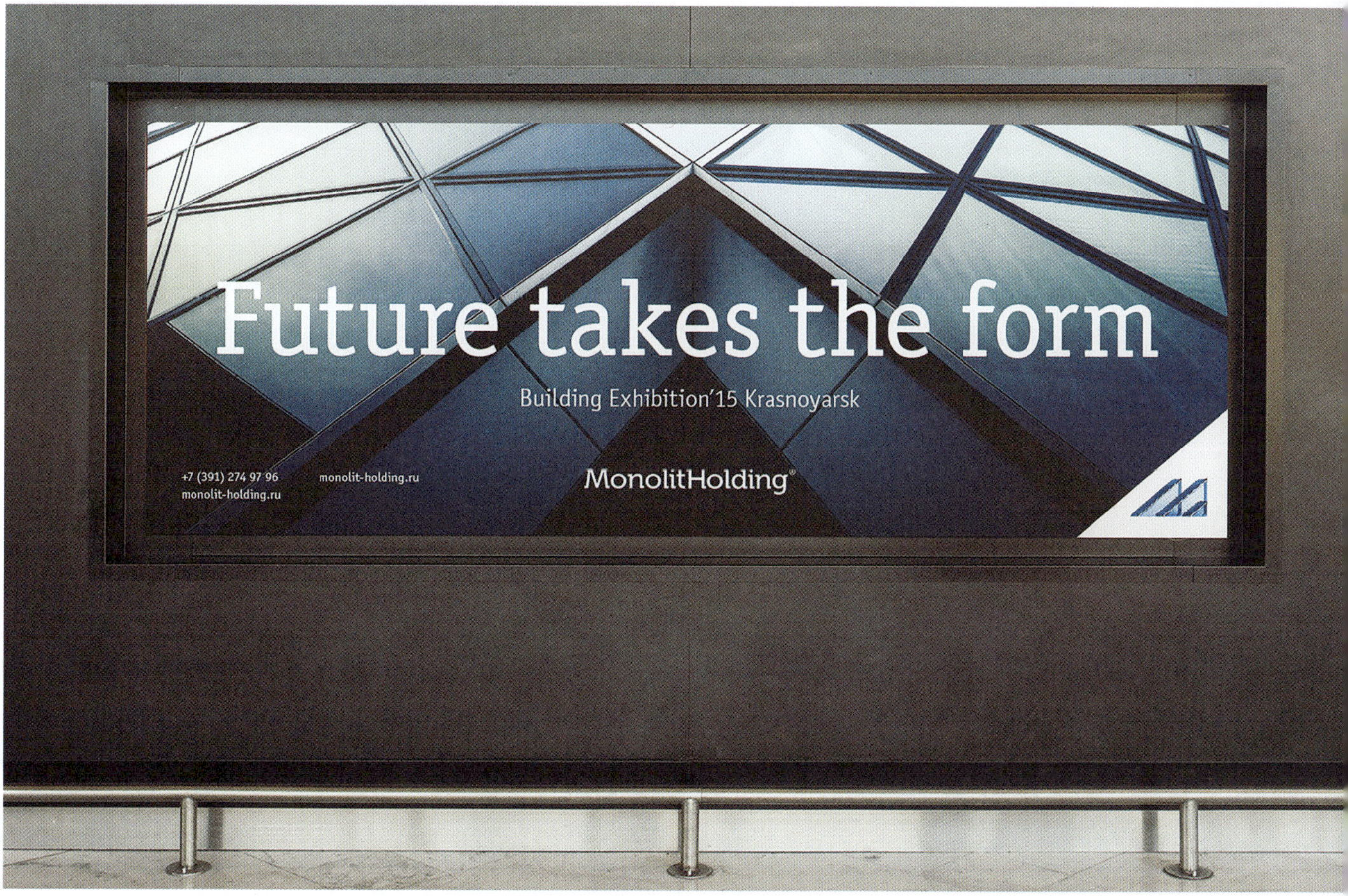

+7 (391) 274 97 96
monolit-holding.ru

Future takes the form

Building Exhibition'15 Krasnoyarsk

+7 (391) 274 97 96
monolit-holding.ru   monolit-holding.ru

MonolitHolding®

TITLE

MUTUO

CREDITS

Studio — **FUTURA**
Photographer — **Caroga**

DESCRIPTION

Mutuo, two architects proposing solutions that represent an ode to the experimentation. Taking advantage of their different roots, they create new experiences through chaos.

This project allowed us to came out with a particular solution: an identity with no logo, that relies in other resources to accomplish the characteristics of recognition and constancy that branding requires.

SVEGLIO

CREDITS

Studio — **Menta Picante**
Creative Director — **Alejandro Román**
Project Director — **Gabriela Salazar**

DESCRIPTION

Project that involves the graphic identity for an architectural firm.

Sveglio is a firm of architects specialized in the construction and detailed of buildings characterized for using luxury finishes. They have strategic alliances with exclusive suppliers of Guadalajara.

a_31 Reid Street, Ballarat,
Victoria 3350.
m_0423 016 661
e_hello@approvalpartners.com.au

w_approvalpartners.com.au

Twitter @approvalpartners
Linkedin/approvalpartners

Approval
Partners

Twitter @approvalpartners
Linkedin/approvalpartners

Approval
Partners

a_31 Reid Street, Ballarat,
Victoria 3350.
m_0423 016 661
e_hello@approvalpartners.com.au

w_approvalpartners.com.au

Twitter @approvalpartners
Linkedin/approvalpartners

Approval
Partners

a_31 Reid Street, Ballarat,
Victoria 3350.
m_0423 016 661
e_hello@approvalpartners.com.au
w_approvalpartners.com.au

Twitter @approvalpartners
Linkedin/approvalpartners

Approval
Partners

Twitter @approvalpartners
Linkedin/approvalpartners

a_31 Reid Street, Ballarat,
Victoria 3350.
m_0423 016 661
e_hello@approvalpartners.com.au

w_approvalpartners.com.au

Twitter @approvalpartners
Linkedin/approvalpartners

a_31 Reid Street, Ballarat,
Victoria 3350.
m_0423 016 661
e_hello@approvalpartners.com.au

w_approvalpartners.com.au

Twitter @approvalpartners
Linkedin/approvalpartners

Approval
Partners

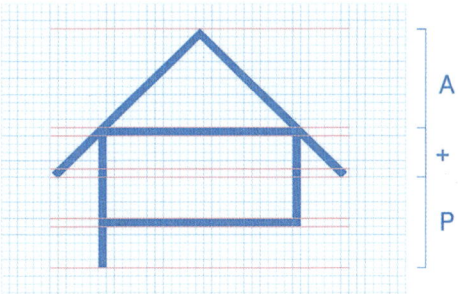

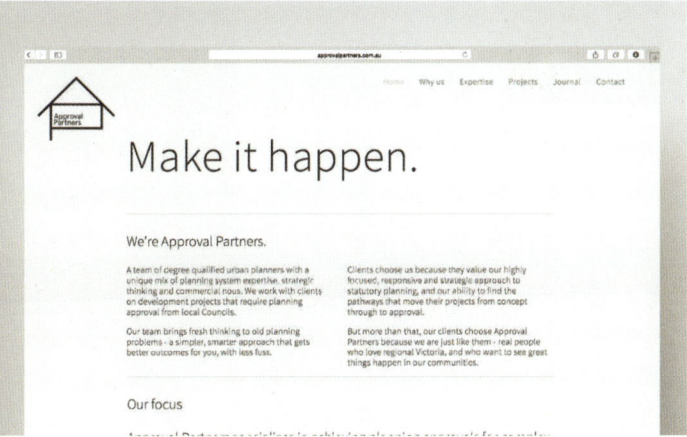

APPROVAL PARTNERS

Studio — **makebardo**

Approval Partners is a team of degree qualified urban planners with a unique mix of planning system expertise, strategic thinking and commercial nous. Their passion is guiding client projects through the statutory approvals process as efficiently and smoothly as possible. They started life as part of URBANSCALE, working with clients on development projects that required new ways of thinking in order to resolve historical planning issues.

From these experiences they learned that clients valued their highly focused, responsive and unconventional approach to planning, and their ability to find the strategic and political linkages that progressed projects from the initial concept stage through to approval. They have created a new team comprising of regionally-based planning experts attuned to the commercial realities of projects and an excellent grasp on local policy, politics and specific spatial factors.

Our challenge was to create a new brand identity distinctive from every other planning consultant. The main words that our client give us were "sharp, focussed, friendly, confidence, certainty, expertise." They wanted AP in capitals in some of the branding but were very open and approachable with a light touch allowing us to develop a fresh concept.

We produced a clean and refined logotype and colour palette that are sympathetic to the minimalist qualities of the brand soul. We achieved an intelligent solution comprising the logotype with capital letters in each word of the name: AP. These letters build a unique symbol of "home," being representative of their field of expertise. Our response was to create a solution that clearly identified the soul of the brand, with recognisability and integrity.

Approval Partners

To Whom It May Concern:,

Dolor sit amet, consectetur adipisicing elit, sed do eiusmod tempor incididunt ut labore et dolore magna aliqua. Ut enim ad minim veniam, quis nostrud exercitation ullamco laboris nisi ut aliquip ex ea commodo consequat. Duis aute irure dolor in reprehenderit in voluptate velit esse cillum dolore eu fugiat nulla pariatur. Excepteur sint occaecat cupidatat non proident, sunt in culpa qui officia deserunt mollit anim id est laborum.

Sed ut perspiciatis unde omnis iste natus error sit voluptatem accusantium doloremque laudantium, totam rem aperiam, eaque ipsa quae ab illo inventore veritatis et quasi architecto beatae vitae dicta sunt explicabo. Nemo enim ipsam voluptatem quia voluptas sit aspernatur aut odit aut fugit, sed quia consequuntur magni dolores eos qui ratione voluptatem sequi nesciunt. Neque porro quisquam est, qui dolorem ipsum quia dolor sit amet, consectetur, adipisci velit, sed quia non numquam eius modi tempora incidunt ut labore et dolore magnam aliquam quaerat voluptatem. Ut enim ad minima veniam, quis nostrum exercitationem ullam corporis suscipit laboriosam, nisi ut aliquid ex ea commodi consequatur? Quis autem vel eum iure reprehenderit qui in ea voluptate velit esse quam nihil molestiae consequatur, vel illum qui dolorem eum fugiat quo voluptas nulla pariatur.

But I must explain to you how all this mistaken idea of denouncing pleasure and praising pain was born and I will give you a complete account of the system, and expound the actual teachings of the great explorer of master-builder of human happiness. No one rejects, dislikes, or itself, because it is pleasure, but because those who do not know pleasure rationally encounter consequences that are extrem again is there anyone who loves or pursues or desires to obtain

Kind regards,

a_31 Reid Street, Ballarat,
Victoria 3350,
m_0423 016 661
e_hello@approvalpartners.com.au
w_approvalpartners.com.au

Twitter @approvalpartners
LinkedIn/approvalpartners

a_31 Reid Street, Ballarat,
Victoria 3350,
m_0423 016 661
e_hello@approvalpartners.com.au

w_approvalpartners.com.au

Approval Partners

Twitter @approvalpartners
LinkedIn/approvalpartners

a_31 Reid Street, Ballarat,
Victoria 3350,
m_0423 016 661
e_hello@approvalpartners.com.au

w_approvalpartners.com.au

Twitter @approvalpartners
LinkedIn/approvalpartners

Approval Partners

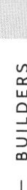

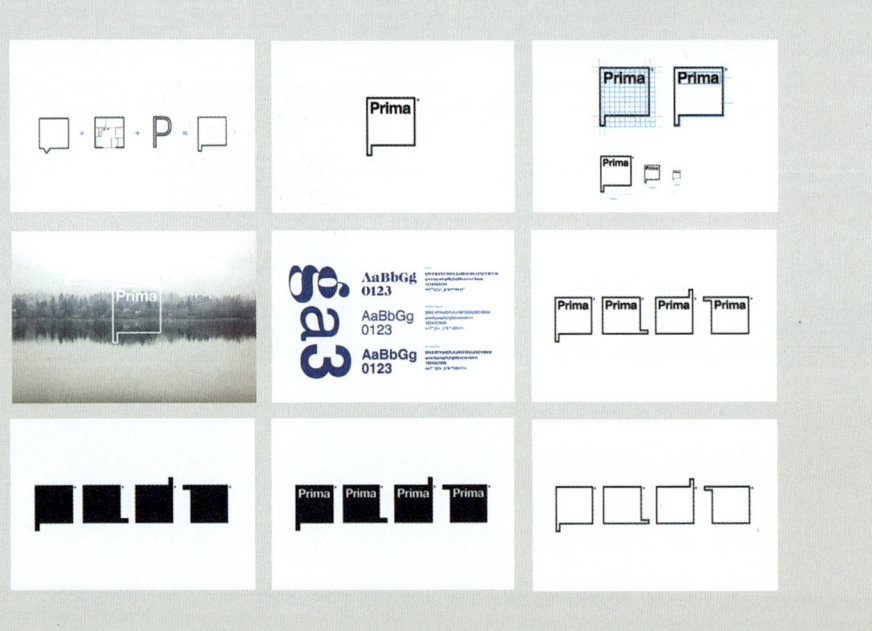

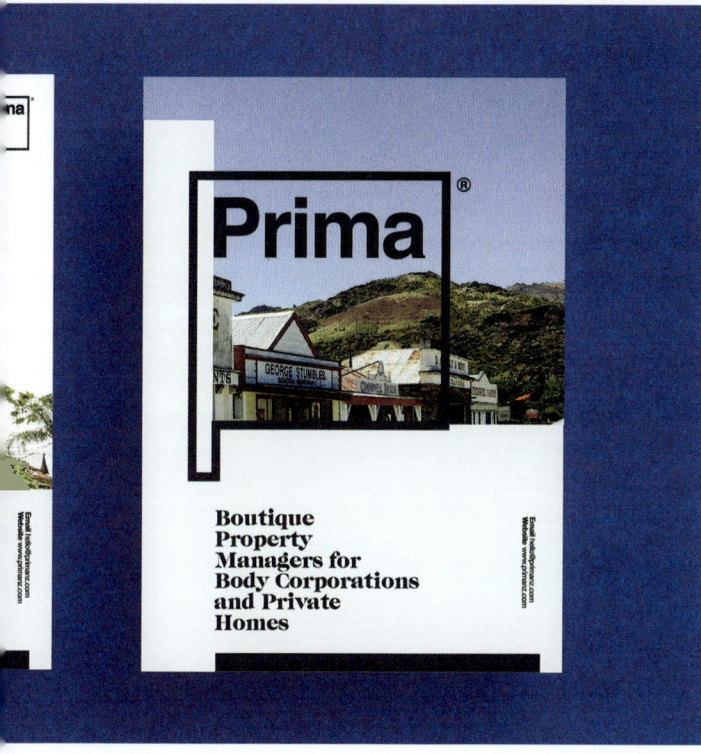

PRIMA

Studio — **makebardo**

DESCRIPTION

Prima is headed up by Sara & Alasdair Roy, with over 30 years Queenstown property experience between them. Sara previously managed Jack's Points Home Owners Association and Prima is proud to include this Association as a client.

For this project we worked with values such as quality, professionalism, experience, discretion, trust and reliability. These values perfectly describe the professional and human characteristics of Sara & Alasdair Roy. Also a key factor was to show that they work as a service boutique company, meeting the needs of their customers in a personal way.

Through a collective approach, the design team listened and gathered important contributions from Sara and Alasdair about their requirements and feeling in their business. We believe that the idea of "boutique" is not about scale, but about proximity to clients. One key factor is fluid communication with their clients and taking care of the space that they manage as though it is their home.

The use of the bright blue colour is associated with loyalty and honesty. It is a spiritual, fresh and transparent colour. Our petrol blue is a colour that conveys maturity and wisdom. This colour suggests responsibility, loyalty and inspires confidence. The combination of both these colours brings a perfect equilibrium, identifying the soul of the brand.

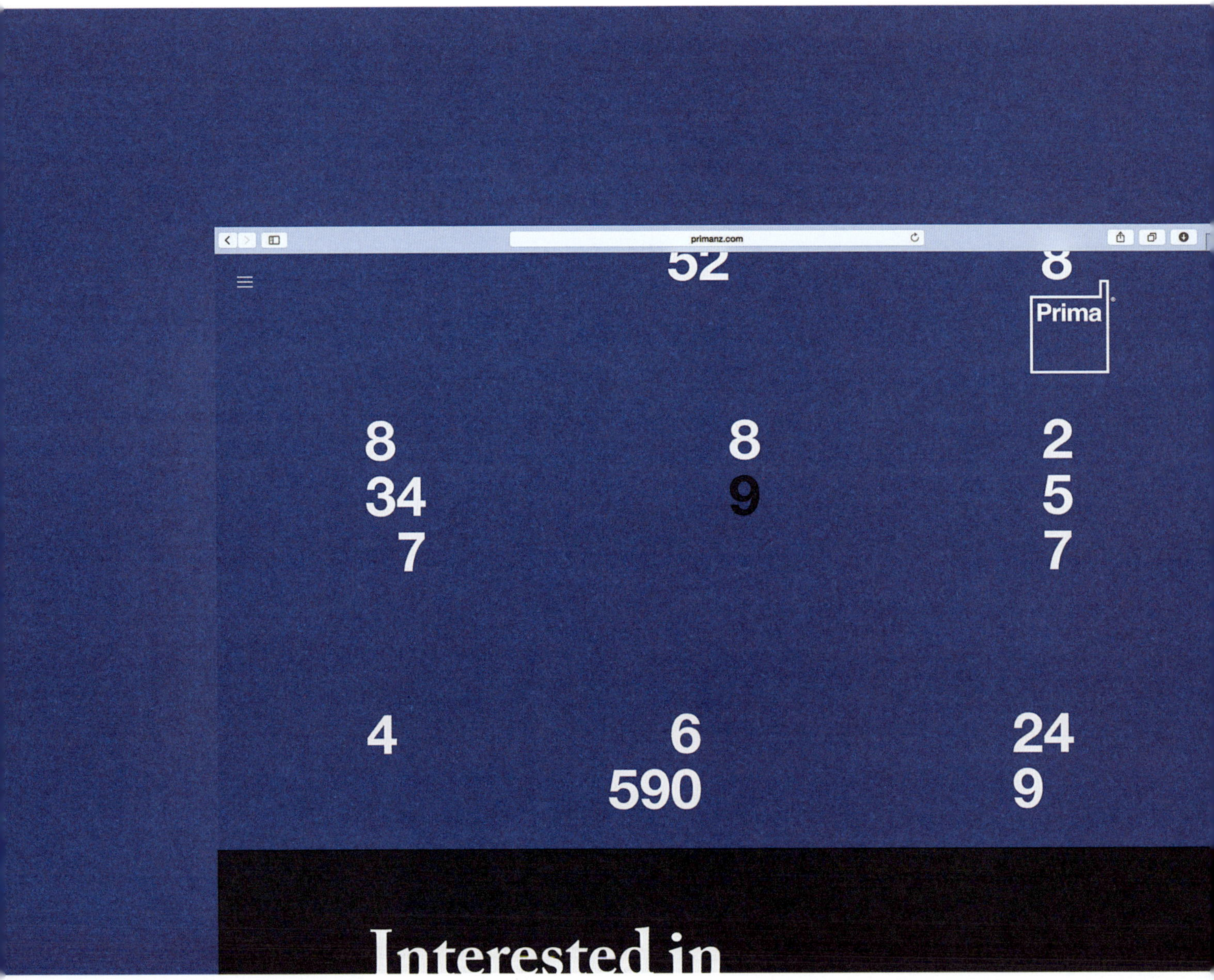

primanz.com

52

8

≡

Prima®

8
34
7

8
9

2
5
7

4

6
590

24
9

Interested in

FRANCELINO IMÓVEIS

CREDITS

Studio — **Jazz Design**
Designer — **Petter Martins**

DESCRIPTION

After completing 45 years of history, the Francelino Imóveis decided to reinvent itself. Underwent internal restructuring processes, change management and team and also changed its market performance.

Formerly known as a great builder, contributed greatly to the growth of neighbourhoods in the city of Itajaí — Santa Catarina State — and now want to regain the value of your brand in the market, becoming reference in the sale and rental of residential and commercial properties in city.

The redesign of its visual identity shows to the public the desire to be an ever new company, with a great ally in its competitiveness strategy.

Together with the client, and precisely because of its tradition, we decided that we should keep a heritage of the old brand, which is your initial designed from a perspective. The heritage is also kept in blue colour as an important part of the visual identity, where only we adapt the old tone for the appropriate.

francelinoimoveis.com.br

Compre seu primeiro apartamento.
francelinoimoveis.com.br

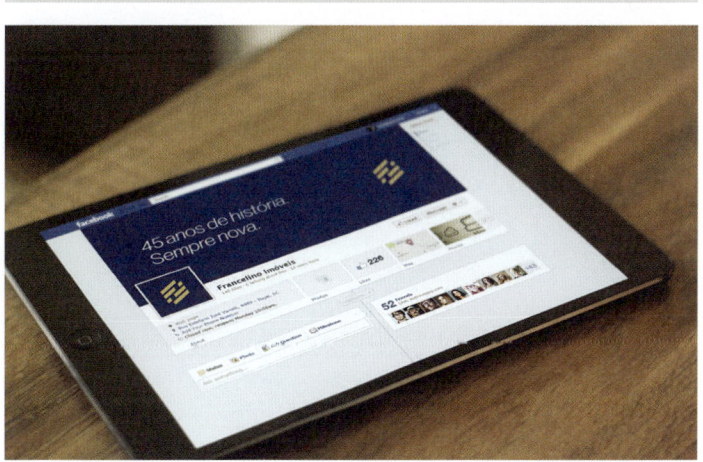

BUILDINGS

STRATA

CREDITS

Studio — **Blast Design Ltd.**
Creative Director — **Colin Gifford**
Designer — **Nick Smart**
Client — **LaSalle Investments**

DESCRIPTION

Naming, brand identity and marketing communications for "Strata," a landmark commercial property development in Staines upon Thames. The sleek, three-tiered structure of the building is echoed in every strand of the identity through the use of colour, layering of information and the three key messages: "Unique, Design, Advantage."

All pre- and post- completion marketing materials were created to reflect the sophisticated design and high quality specification of the building. The campaign was launched with a limited edition personalised box, which featured layers of information printed, embossed, foil blocked and laser etched into different materials.

strata
Unique
Design
Advantage
Floor Plans
First floor
Floor Plans
Second floor

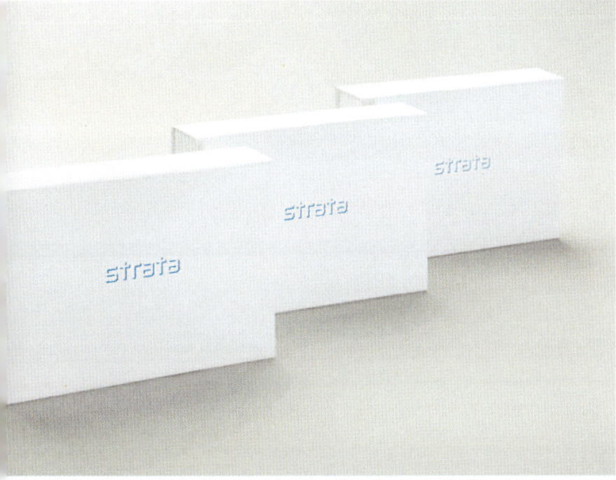
strata
strata
strata

stratastaines.com

1.0

2.0

2.1

2.2

3.1

stratastaines.com

Strata Staines

3.0

o Connectivity
o Proximity
o Status

MALLORCA

CREDITS

Studio — **Menta Picante**
Project Director — **Gabriela Salazar**
Creative Director — **Alejandro Román**

DESCRIPTION

Project of graphic identity and
brochure for the development of luxury
residential towers in Guadalajara.

The skyscrapers Mallorca, will be located
in one of the greatest areas of Guadalajara,
being in Providencia offer the highest
standards of quality of life. It is a modern,
exclusive and luxury development.

MANTA

CREDITS

Studio — **THERE**
Designer — **Paul Tabouré, Jon Zhu**
Photographer — **THERE**
Client — **McGrath**

DESCRIPTION

Property expert John McGrath of McGrath Real Estate Agents, appointed us to create the branding and marketing of Manta – a brand new residential development in coastal Wollongong, aimed primarily at young professionals, couples and empty nesters.

The brand campaign centered around positioning Manta as offering an affordable yet contemporary relaxed coastal lifestyle. Bright, sun-kissed photography coupled with a distinctive hand written text style, presented a very fresh, personable and emotive image that perfectly captured the appealing blend of indoor and outdoor living available on the NSW coast.

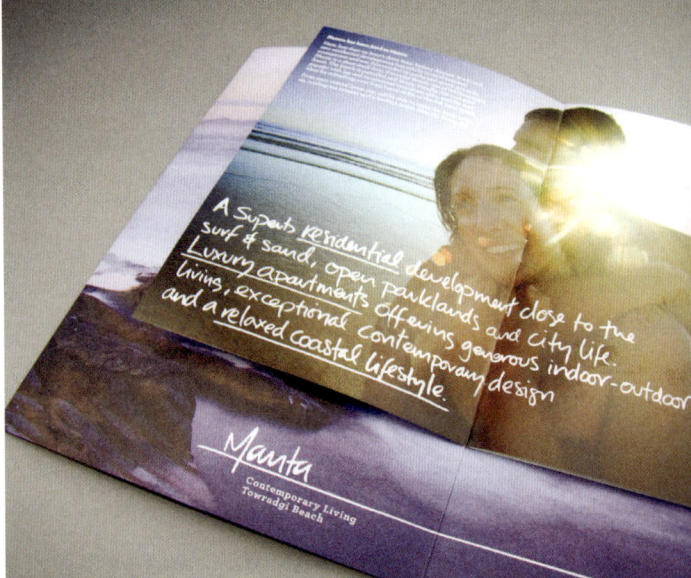

1 min to the surf
5 mins to downtown
45 mins to the Shire
60 mins to the airport

Sydney

Sydney Airport

Botany Bay

Princes Hwy

Sutherland

Royal National Park

Imagine beginning your day with a short
stroll down to the white sand, rolling surf
and rock pool. Whether you want to swim in
the pool, ride a long board or watch the sun
rise out of the ocean, it is an invigorating
start to any day.

Enjoy a long, slow breakfast at your on-site
cafe. If you need to venture further afield,
grab a coffee on the run. Head into the city,
or to the park.

As the sun drops low over the escarpment,
share a story at the communal barbeque area,
pop into town for dinner or simply kick back
on your balcony. Room for all.

Then tomorrow begins at Manta,
another day rich in possibility.

Celebrate the everyday

Manta
Contemporary Living
Towradgi Beach

www.mantaliving.com.au

THE CAPITOL

Studio — **FABIO ONGARATO DESIGN**
Creative Director — **Fabio Ongarato**
Collaborators — **Jean-Philippe Delhomme,
Filipe Jardim, Daniel Riera, Sebastian Gollings,
Ben Briand, Cameron Bruhn, Bates Smart**

The Capitol is the premier South Yarra luxury
apartment address for Melbourne's urban
elite, on the corner of Toorak Road and Chapel
Street. Designed by Bates Smart, The Capitol
offers unrivalled designer appointments and
generous spaces; setting a new benchmark
for premium apartment living.

For this project we created "The Capitol" brand and
developed a highly integrated marketing campaign,
employing the collaboration of international artists
and photographers to deliver a unique approach
that eschews the typical property development
promotion, and depicts "the Capitol way of life."

ON SOCIETY & CULTURE

ON SERVICE

ON ATMOSPHERE

PALAIS PRINCIPE

CREDITS

Studio — **moodley brand identity**
Client — **JP Immobilien**
Creative & Art Director — **Gerd Schicketanz**
Graphic Designer — **Nora Obergeschwandner**, **Gerald Schinwald**
Text & Book Concepts — **Michael Endlicher**
Illustrators — **Daniel Egnéus**

DESCRIPTION

moodley brand identity has developed extraordinary commercial visualizations, "branded addresses," for the Viennese real estate project Palais Principe — luxury high-end apartments in a revitalized building from the Art Nouveau period. In order to communicate the building's own speciality and identity a special image booklet, written by an imaginary resident of the Palais in the style of a diary, explains the history of the Palais and its special urban setting. The "diary" written in handwriting also includes exquisite illustrations done by an artist.
The logo is put on the hardcover of the high quality image book by means of hot foil stamping.

FUTURIS

Studio — **Vide Infra**
Creative Director — **Alexey Draganov**

Futuris is a new residential building in the very heart of Riga. We were assigned to create a brand and a marketing website that would help the building stand out with a unique and authentic message.

FUTURIS
MODERN LIVING

FUTURIS is a dwelling house with 70 flats, built using the latest technologies. Owing to the well-appointed architecture, rational layout solutions, a great location, and an excellent surrounding infrastructure, FUTURIS provides you with a harmonious living environment and fully matches the contemporary standards of a comfortable life.

FUTURIS

ANTONIJAS IELA 16 A
RĪGA

FUTURIS

ANTONIJAS IELA 16 A
RĪGA

FUTURIS
MODERN LIVING

FUTURIS

ANTONIJAS IELA 16 A
RĪGA

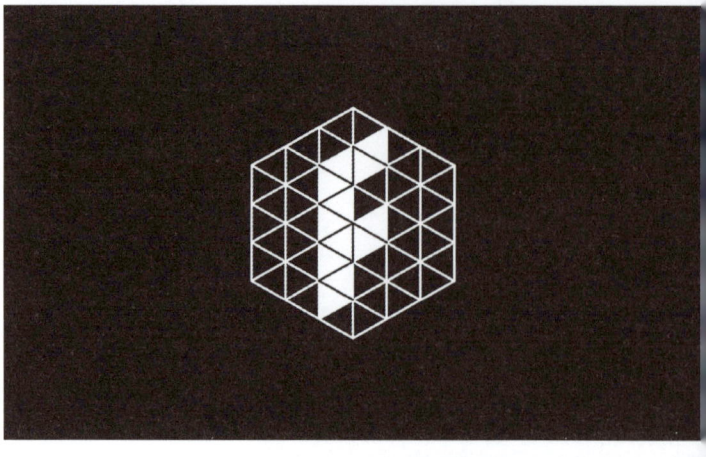

FORBURY PLACE

CREDITS

Studio — **Blast Design Ltd.**
Creative Director — **Colin Gifford**
Designer — **Jamie Conkleton**
Client — **M&G Real Estate**
www.forburyplace.com
Printer — **Push**

DESCRIPTION

Naming, brand identity and marketing
communications for Forbury Place: the pre-
eminent business destination in the South East.
Architects Aukett Fitzroy Robinson's concept
was to create façades of the building that appear
different from every angle. With their vision
in mind, we developed our marketing concept
"however you view it" to communicate the
business and lifestyle benefits, drawing from
the buildings' distinctive structural "diagrid"
for the design of the logo and identity.

The scheme is the largest and most prestigious
development to be undertaken in the region,
and the quality of the marketing materials had
to match these aspirations. We created a teaser
campaign featuring 3D laser-etched glass cubes
and an animated film. This was followed by a
website presenting films that included interviews
with the investor, developer and architect.
The scheme was launched with a
screening of two films at the Soho Hotel
and the launch of a hardback book.

ANTHOLOGY

Studio — **Greenspace**
Founder & CEO — **Adrian Caddy**
Creative Director — **Lee Deverill**
MD & Strategy Director — **Lene Nielsen**
Designer — **Jamie Scott, Nicole Smith,
Andrei Verioti**
Senior Account Manager — **Mirrelle Rank**
Account Manager — **Gabriela Sperling**

Often the most exciting projects to work on
are those that start with a blank sheet of
paper — and when that is combined with a
client that has real ambition, expertise and
strong backing, it can come together to create
a brand that makes a real difference.

Anthology is a large-scale residential London
property development company, backed
by Oaktree's European Principal Fund. Its
strategy is very simple, to build homes in
London, for Londoners. In an industry where
brands are not always respected, Anthology
will focus on marketing and service excellence
to help guide customers and to inspire
stories of homes, built from London.

In addition to an extensive strategic and naming
process, Greenspace created a master brand
and visual identity system that will evolve with
each new development, alongside marketing
collateral and bespoke on-line tools.

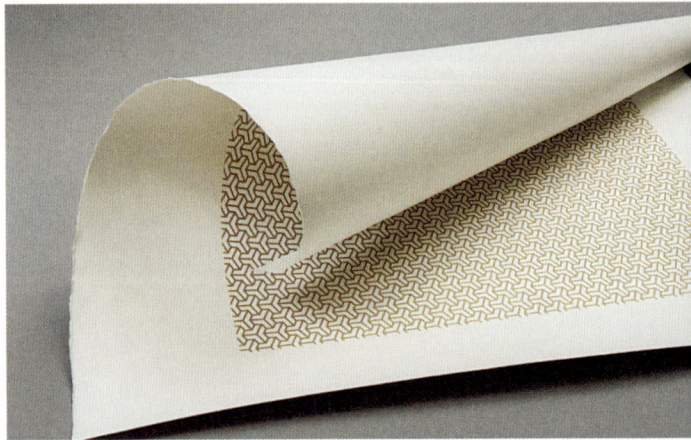

HENRY VIII FOUNDED DEPTFORD DOCKYARD IN 1513

Deptford first became a settlement of national importance during the Tudor period. The Deptford Dockyard was used to build and maintain warships for 350 years.

The most famous of the Elizabethan sea-heroes is Sir Francis Drake who became the second person ever to circumnavigate the world, aboard his ship the Golden Hind.

Drake cemented England as a serious rival to Spain in the age of discovery, and to acknowledge this, Queen Elizabeth I went to the dockyard in 1581 to award him with a knighthood aboard his ship.

So many people clambered across the wooden dockyard bridge to see the Queen that it broke, with more than a hundred people falling into the river, though no-one was hurt.

PARKLANDS

Studio — **THERE**
Designer — **Paul Tabouré, Jon Zhu**
Photographer — **THERE**
Client — **Chase Property**

DESCRIPTION

Parklands is a luxury residential development
located in the heart of the Blue Mountains,
NSW and nestled amongst a breathtaking 28
acre parkland landscape of undulating lawns,
manicured gardens and towering 100yr old trees.

This extraordinary development by Chase
Property, Nettleton Tribe Architects and
Cramer Property Sales, comprised 64 luxurious
2 and 3 bedroom semi-attached houses, each
occupying its own exquisite landscape.

From the beautifully crafted development
identity and typography, to the meticulously
art directed imagery and poetic copy, THERE
created a brand campaign which captured
the essence of this magical place where
you could be forgiven for thinking you had
arrived at an aristocratic European manor.

The successful campaign helped deliver
over $3million worth or pre-sales in the
first weekend alone and eventually led to all
residences being sold ahead of schedule.

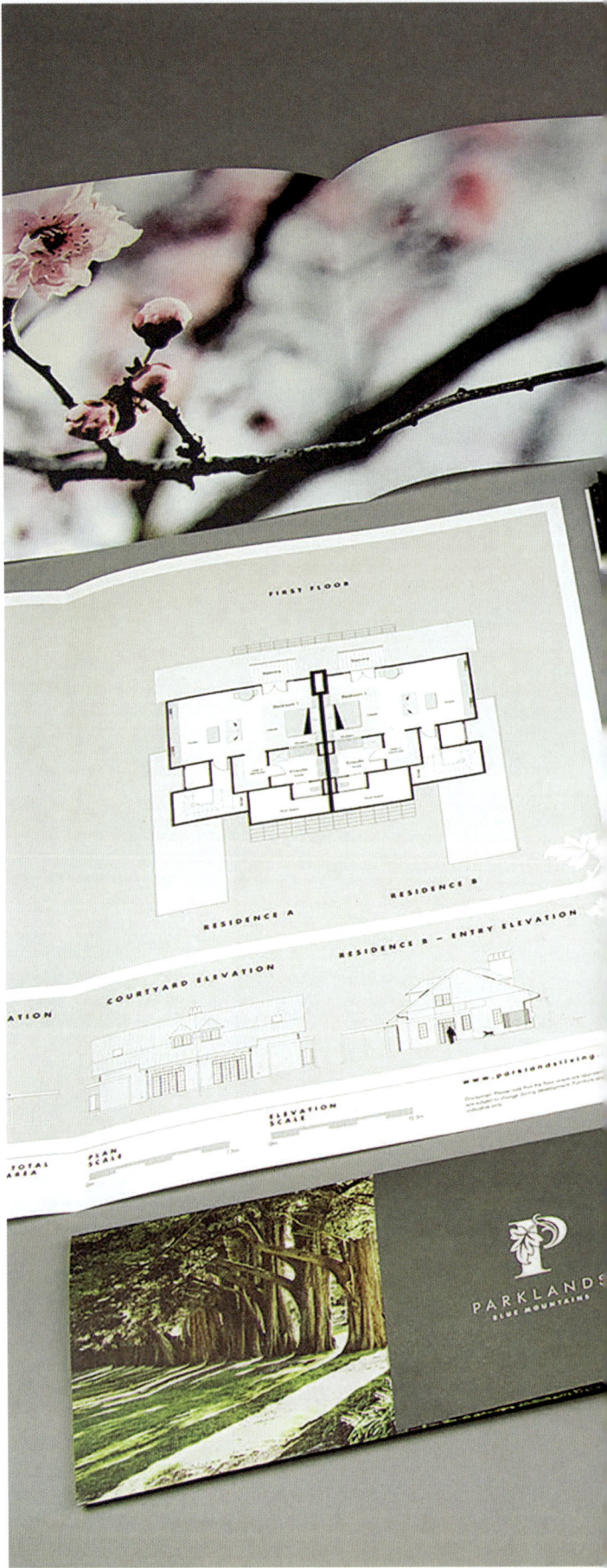

Where an enchanting lake lies tranquil, surrounded by acres of pristine lawn & giant arbors...

PARKLANDS
BLUE MOUNTAINS

LEVEL

CREDITS

Studio — **Vide Infra**
Creative Director — **Alexey Draganov**

DESCRIPTION

Find a balance in your life! Level Residence
offers a balanced living in a comfort of a city
apartment surrounded by a natural landscape
of Moscow region. The website explores this
balance with a rich responsive presentation and
a collection of immersive interactive features.

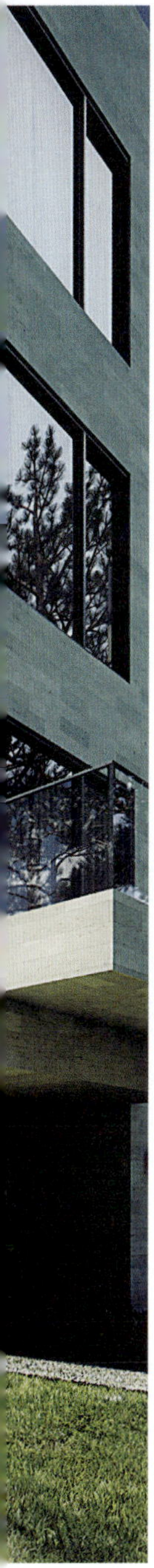

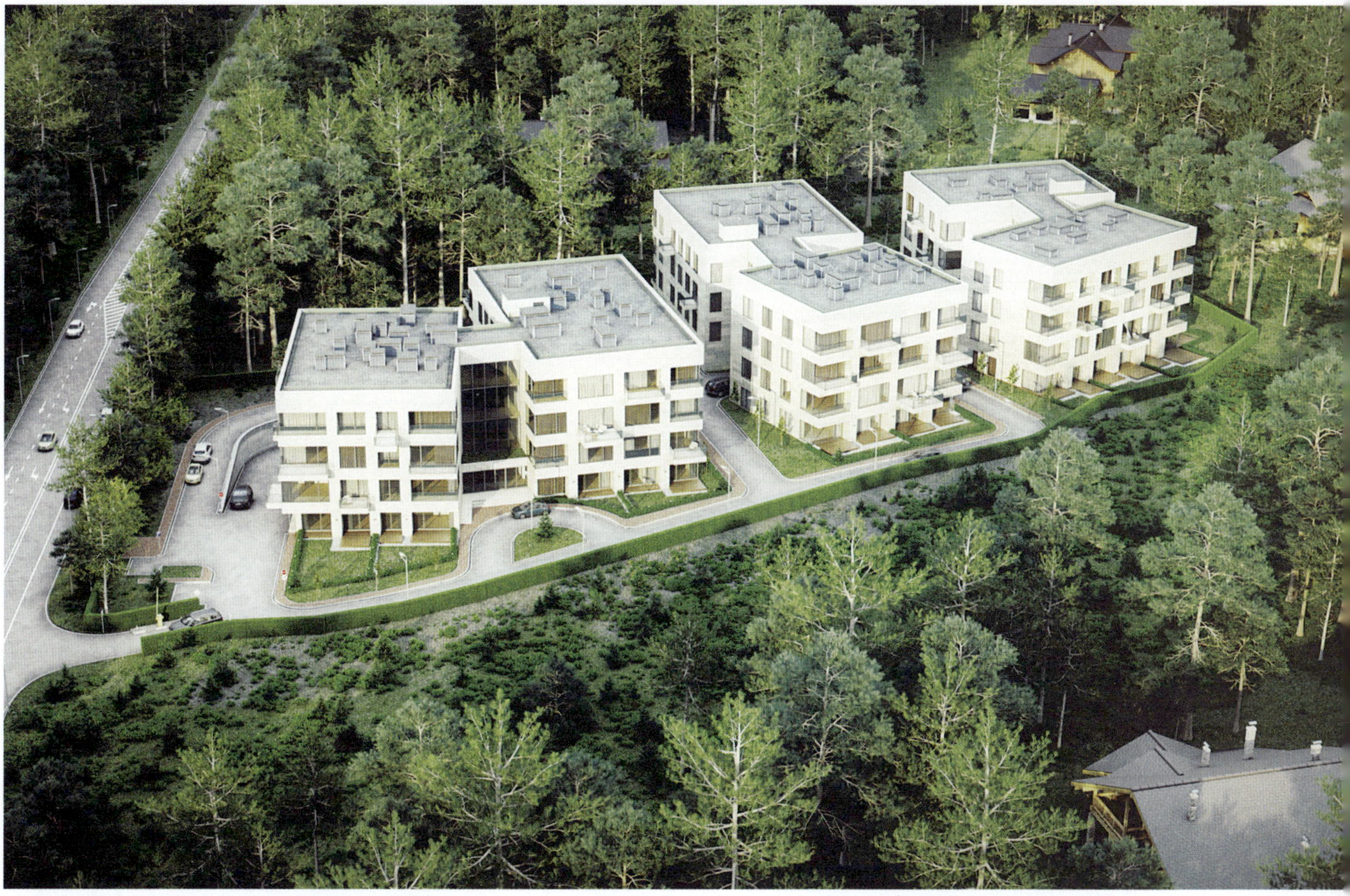

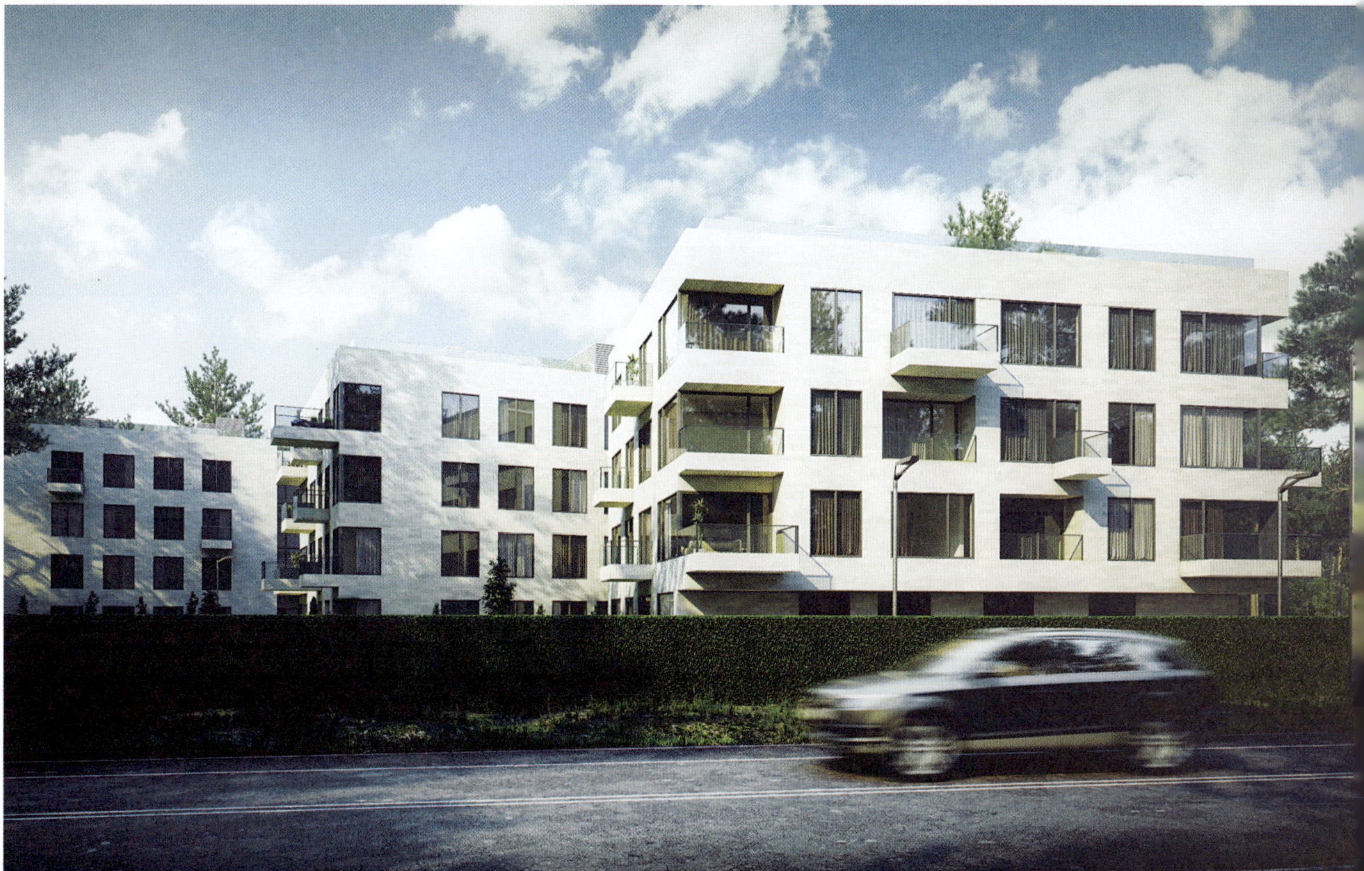

DÉCIMA

CREDITS

Studio — **FIRMALT**
Creative Director — **Manuel Llaguno**
Lead Designer — **Déborah G. Neaves**

DESCRIPTION

Décima is a six story complex of corporate offices.
It offers extravagant spaces for businesses to grow
and develop. It is strategically located within the
city for its easy access, and has privileged views
of the majestic mountains that surround the city.

The branding proposal comes from a sense of
belonging, which is portrayed in Décima's emblem.
The shield serves as a reminder for the people who
work, buy, or rent their office there, that they can
grow and transcend within Décima's community.

The brand's aesthetic is built around heritage
and elegance, adding modern twists throughout
the applications. Décima's typographic system
irradiates tradition, while icons like wheat stalks,
flags, and a rising sun maintain the brand's
contemporary approach. Each icon has a reason
to be — the wheat stalks make reference to the
territory, the flag represents establishment in such
territory, and the rising sun is the new beginning.

Décima's colour palette plays along with the
traditional concept of the brand, entertaining
colours like brown, green and ivory. The copper
works as a tribute to the concept of business
exchange, as it is known to be used in the coinage
of currency. The foil is applied throughout the
collateral, highlighting Décima's prosper essence.

Décima

Décima

BENITO JUÁREZ 898 ZONA DE LOS CALLEJONES SP GG, NL

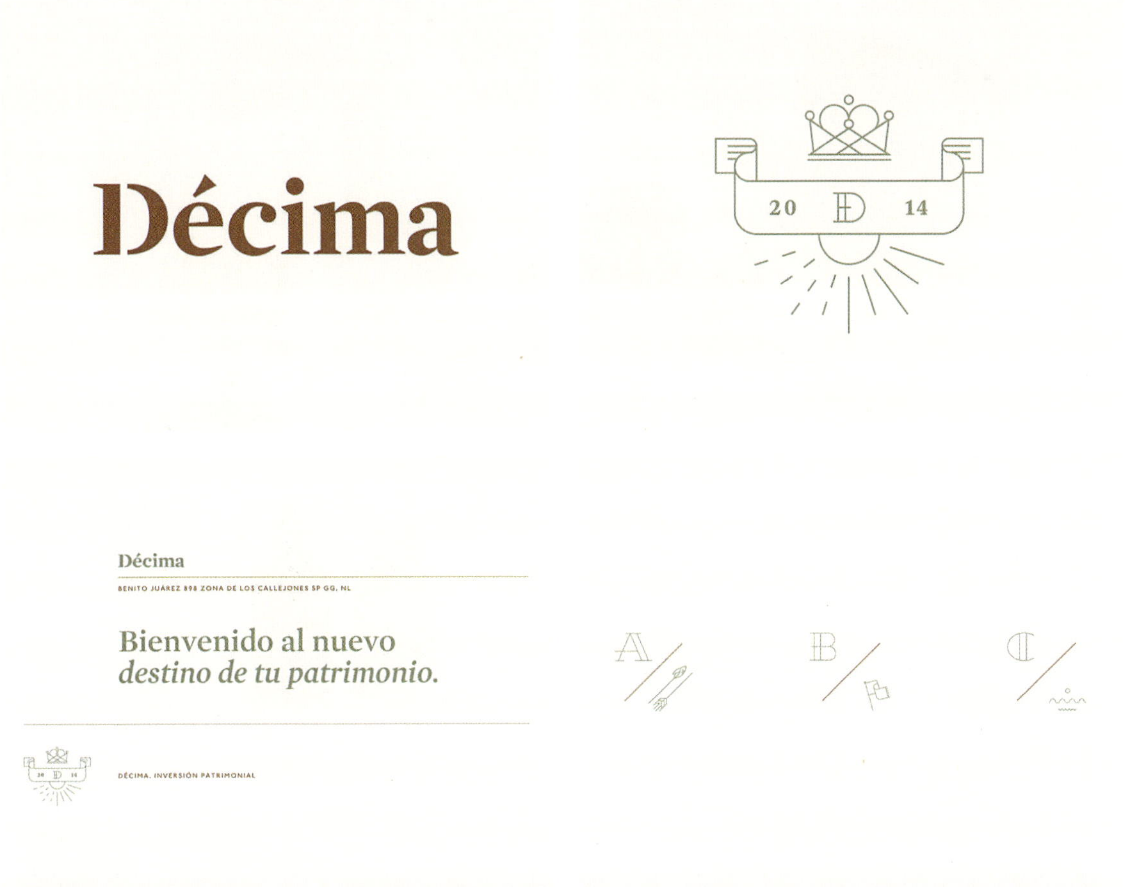

Bienvenido al nuevo
destino de tu patrimonio.

DÉCIMA. INVERSIÓN PATRIMONIAL

PLANTA BAJA
DÉCIMA. INVERSIÓN PATRIMONIAL

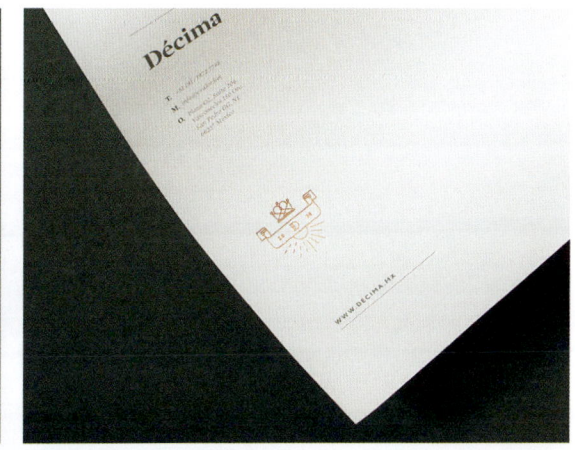

MINTON HOUSE

CREDITS

Studio — **THERE**
Designer — **Paul Tabouré, Jon Zhu**
Photographer — **THERE**
Client — **Hayson Group**

DESCRIPTION

Minton House is an iconic art deco building located at the gateway to Kings Cross and beautifully restored by Hayson Group with Marchese & Partners.

This commercial use development comprising over thirty studios, boasted a rich creative history with many famous organisations from the film industry, such as Tropfest and Animal Logic, starting up there.

We developed a set of contemporary yet evocative brand images and implemented an integrated campaign across traditional, digital and PR media that captured the essence of 'work studios with a twist' whilst helping portray the building's eccentric and flamboyant personality with a nod to its nostalgic and creative heritage.

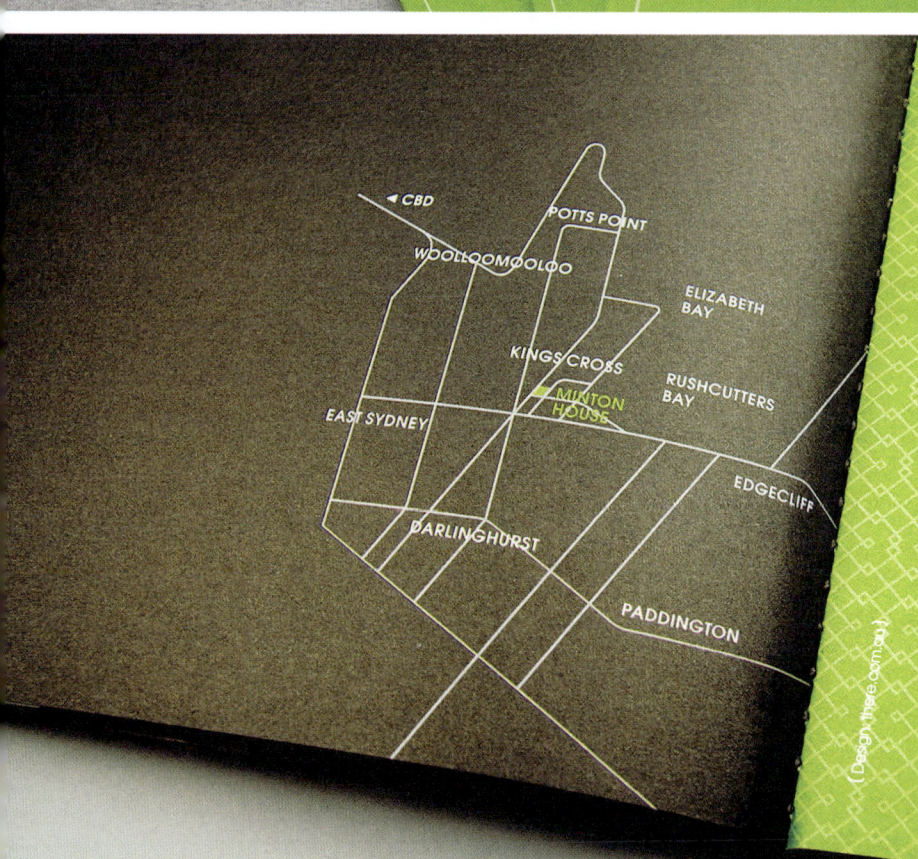

A sneak peek
of a unique
new chapter...

◄ CBD
POTTS POINT
WOOLLOOMOOLOO
ELIZABETH BAY
KINGS CROSS
MINTON HOUSE
RUSHCUTTERS BAY
EAST SYDNEY
EDGECLIFF
DARLINGHURST
PADDINGTON

(Design Here.com.au).

The gateway to a new Kings Cross

Space to blow your own trumpet

If the walls could talk

Oscar-quality investment

DULWICH GREEN

Studio — **THERE**
Designer — **Paul Tabouré , Dave Gale**
Photographer — **Steve Brown**
Client — **Haralambis Property**

DESCRIPTION

Haralambis Property is a Sydney based developer with 20 years of design and construction experience. They have a track record in creating high-end award-winning residential developments throughout the inner west.

The Dulwich Green development is located at New Canterbury Road in Dulwich Hill. It is a contemporary development made up of two main buildings set amongst substantial landscaped gardens, courtyards and trees. The courtyard is surrounded by eighty 1, 2 and 3 bed apartments.

Early client stakeholder workshops unearthed the idea of 'Where green meets grunge', reflecting the nature of the development with its cool, edgy, emerging precinct and proposed lush garden enclave.

THERE created a stand-out identity and brand campaign. We created all the required visual assets; local area and lifestyle photography, local maps, custom iconography, CGI consultancy and stylised site and floor plans. Website, brochure, signage, advertising and display suite were produced to aid in selling over $20m+ in the immediate days post-launch.

This is the fourth successful project that Haralambis, Candalepas Architects and THERE have brought to market.

80 DESIGNER APARTMENTS LOCATED IN THE EMERGING, CONNECTED & VIBRANT DULWICH HILL.

BE PART OF SOMETHING NEW AND OLD

Dulwich Hill is a village made up of residents from diverse cultures with a strong sense of community.

While many people have been living here for decades, Dulwich Hill is experiencing a new wave of young families and professionals moving to Dulwich Hill from all over Sydney.

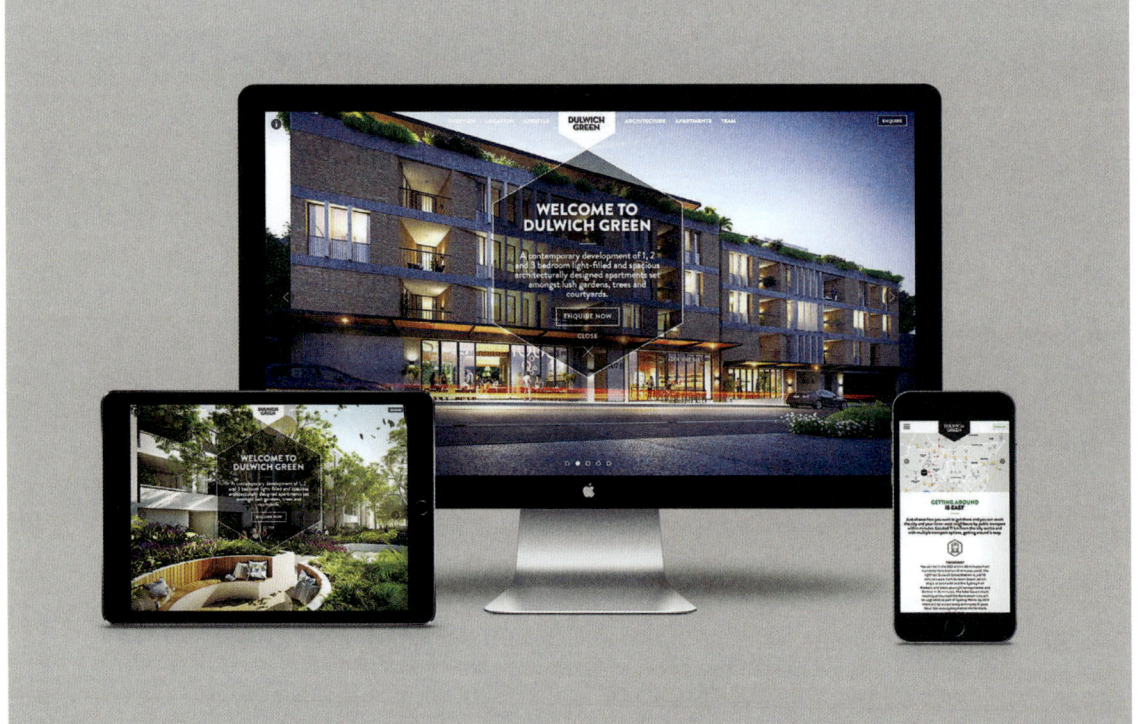

THE BEST THE INNER WEST HAS TO OFFER

GETTING AROUND IS EASY

Dulwic

BE PART OF SOMETHING NEW AND OLD

WELCOME DULWICH GREEN

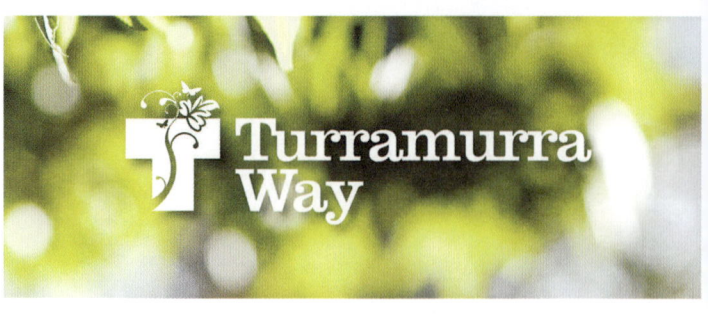

TURRAMURRA WAY

CREDITS

Studio — **THERE**
Designer — **Paul Tabouré, Dave Gale**
Photographer — **THERE**
Client — **AVER**

DESCRIPTION

THERE facilitated an engaging name generation workshop with stakeholder groups. The established name, Turramurra Way, spoke metaphorically to both its location, and the laid-back local neighbourhood spirit.

We undertook a brand campaign strategy that saw us define the unique positioning, 'Your Land, Your Family, Your Way - Turramurra Way' by gaining a deep insight into the offer, its features, the attributes and the target audience, along with their lifestyle, mindsets, emotional drivers, barriers and opportunities.

A lifestyle-focused campaign that appealed to a broad purchaser; local-area purchaser, overseas investor and out-of-area purchaser, whilst capturing the nature-feel reflective of being situated on the edge of the national park. Authentic lifestyle imagery helped convey the sense of life within a strong community.

The campaign was hugely successful generating $30m+ value across three auction tranches – as a result of the media campaign there were over 400 site inspections, 161 registered bidders and 100% of lots were sold at an average of 25% above the reserves price.

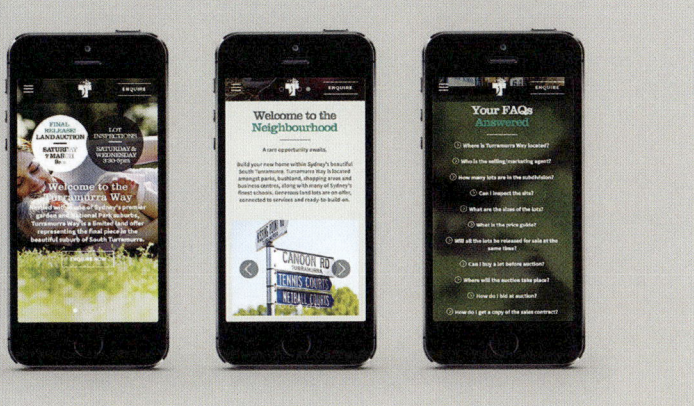

TITLE

COBURG HILL

CREDITS

Designer — **Matt Portch**
Photographer — **Jamie MacFadyen**
Executive Creative Director — **Jason Williams**

DESCRIPTION

Coburg Hill is a modern new development situated close to the city of Melbourne, aimed at late twenties to late thirties. In essence, the brief was to capture the relaxed nature of the surrounding parkland with the vibrancy of being close to the city. The area itself was already "up and coming" with cafés, bars, restaurants and shopping, aimed at a more youthful culture. After developing the logo we found the "Co" in Coburg Hill became a precursor to a series of words used to highlight the feel of the area, i.e. "Cosmopolitan," "Contemporary," "Community," etc. Black was used as the dominant colour to give the premium feel the client wanted. An uncoated stock was used for the brochure to instantiate this and, in places, thin spot varnish pinstripe lines were added to the partition pages of the brochure and stationery. The business cards were kept square to echo the logo itself, as was the goody bag customers took away after a coffee at the sales office.

Coburg Hill

Coburg North's
stylish new community.

Parkland, bike trails and playgrounds make Coburg Hill an idyllic retreat from everyday life.
This stunning development borders the beautiful Edgars Creek Reserve, and is close to Merri Creek,
Jackson's Reserve and Edgars Lake. Whether you're walking, riding or resting, Coburg Hill offers huge
potential for exercise and leisure activities.

Contemporary

Cosmopolitan

Looking for a place to grab a cappuccino or a spot of window shopping? Coburg Hill enjoys
an abundance of local shopping and dining, including the many cafés and restaurants at nearby Sydney
Road and Northland Shopping Centre. Imagine having all this excitement, entertainment and retail therapy
right on your doorstep. There's also a wide range of schools in the area and plans for a brand new
Coburg Hill retail centre.

Introducing Coburg North's stylish new co...
Situated just 9km from the CBD on the histori...
stunning development offers architecturally des...
your own allotment at a highly sought-after add...
is a prime destination in a locality recognised for...
redevelopment and investment potential. Secure ...
Coburg Hill today

PREMIER COLLECTION

CREDITS

Studio — **Journal**
Designer — **Paul Spencer, John-Paul Warner**

DESCRIPTION

Redrow approached design and branding agency Journal to promote their new offer, "Premier Collection" — a range of high-spec penthouses and apartments, positioned across numerous landmark sites in London.

The creative had to sit alongside existing Redrow branding, but also be sufficiently distinct, stylish and elegant in order to attract top-end buyers.

The solution was an interchangeable logo device that can be easily applied to the full range of exclusive Redrow developments. This created a coherent and future proof style for the Premier Collection. The project expanded into an advertising campaign, firstly for Kingston Riverside — a spectacular high-rise complex nestled on the banks of the Thames.

Premier Collection

KINGSTON RIVERSIDE

Premier Collection

ROYAL WATERSIDE

Premier
Collection

AMBERLEY
WATERFRONT

TELEGRAF 7

Studio — **moodley brand identity**
Client — **JP Immobilien**
Creative & Art Director — **Gerd Schicketanz**
Graphic Designer — **Zachary Kutz, Corina Breban**
Text — **Andreas Kump**
Photographer — **Julian Mullan**
Project Manager — **Olivia Forstmayr**

DESCRIPTION

The building with the magnificent façade in
Vienna's 6th district functioned as the telephone
switchboard for the Austria-Hungarian monarchy
from 1899. Today, TELEGRAF 7 is an ideally located
hotspot for the communication sector in mid Vienna
where modern design meets historical architecture.
moodley brand created a branding that perfectly
expresses the exclusive feeling of the property.
Aside from an informative brochure and other
marketing measures such as fact sheets and a
ground plan for the building, an accompanying
website and various print ads in lifestyle and
specialist media have also been created.

2. OBERGESCHOSS

MIETEINHEIT 7

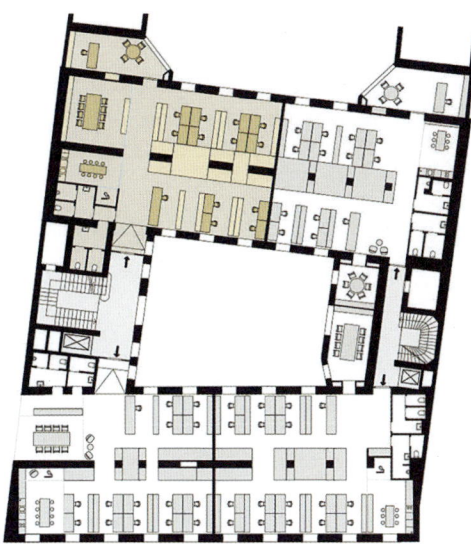

LEHARGASSE

OBJEKTBESCHREIBUNG

MAXIMALKAPAZITÄT
14 ARBEITSPLÄTZE

MIETFLÄCHE/MIETEINHEIT
204,2 m²

MIETFLÄCHE/ARBEITSPLATZ
13,6 m²

≡

NÄCHSTE

TELE
GRAF
7
BÜROHAUS
WIEN

EINE
ABSOLUTE
BESONDERHEIT
DIE INTEGRIERTE HOTELSUITE. ALLE MIETER DES HAUSES KÖNNEN SIE KOSTENLOS NUTZEN, ETWA ZUR EXKLUSIVEN UNTERBRINGUNG AUSWÄRTIGER KUNDEN.
2.
GESCHOSS

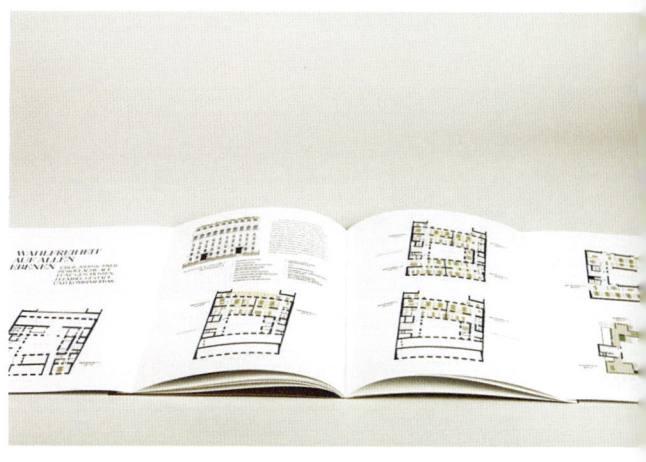

TELE
GRAF 7

BUSINESS
CLASS
AT
WORK

ALBIA

Studio — **SAVVY STUDIO**
Creative Director — **Raúl Salazar**
Account Director — **Orlando Fernández**
Art Director — **Eduardo Hernández**
Designer — **Carolina Larragoity**

DESCRIPTION

Albia is an office building built by Emblem
Capital and situated in the Santa María
corridor in Monterrey, Mexico.

The building is a display of great design by one of
México's most iconic, renowned and prestigious
architects, Agustín Landa Vértiz †. With 5,400
square meters and 150 offices distributed over
19 floors, Albia looks to attract a public that
understands that a modern office should function
not only as an efficient work space, but as an
extension of the personality, philosophy and
nature of the businesses and projects it contains.

Our concept and our basis for the development
of the brand's visual language is the sum of this
philosophy, the façade and the building's raw
materials. The logotype, a symbol for the seven
iconic pillars that visually define and physically
support the building, was designed with simplicity,
elegance and efficiency in mind. These same pillars
are the conceptual overarching element behind
Albia's brand story: seven pillars = seven brand
values: Location, Design, Atmosphere, Distribution,
Functionality, Amenities, and Efficiency.
The colour palette and paper selection are based
on the building's textures and materials, especially
its ample use of concrete and the different
tones it takes through the passage of time.

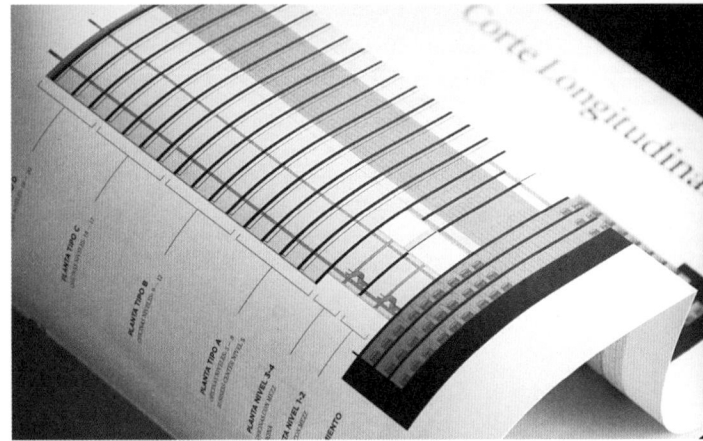

ALBIA
ALBIA.COM.MX

Diseño único por Landa Arquitectos
Distinción y elegancia
Flexibilidad de espacios
AMENIDADES
CARACTERÍSTICAS
ALBIA

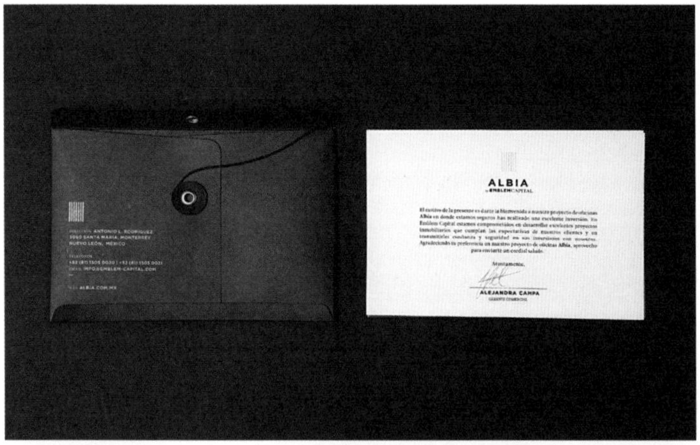

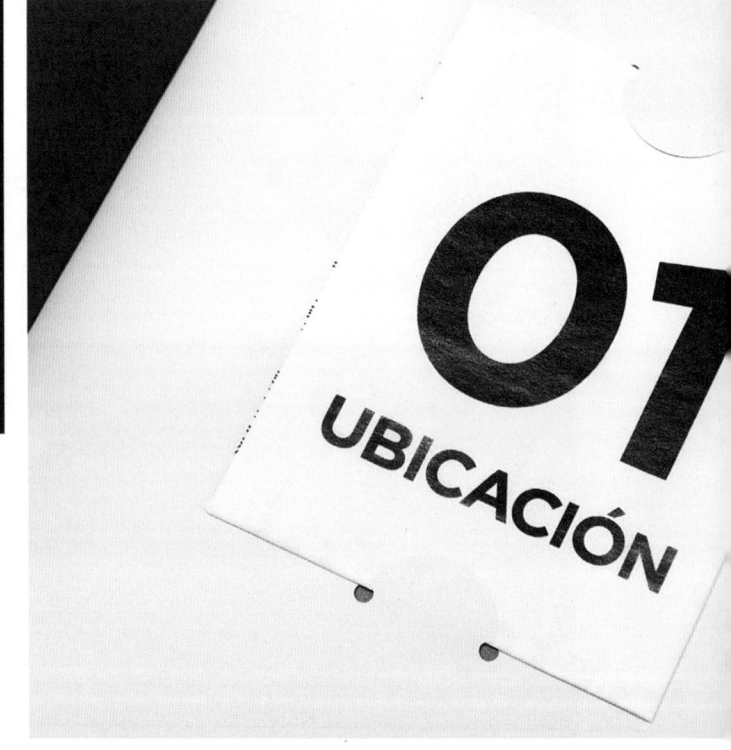

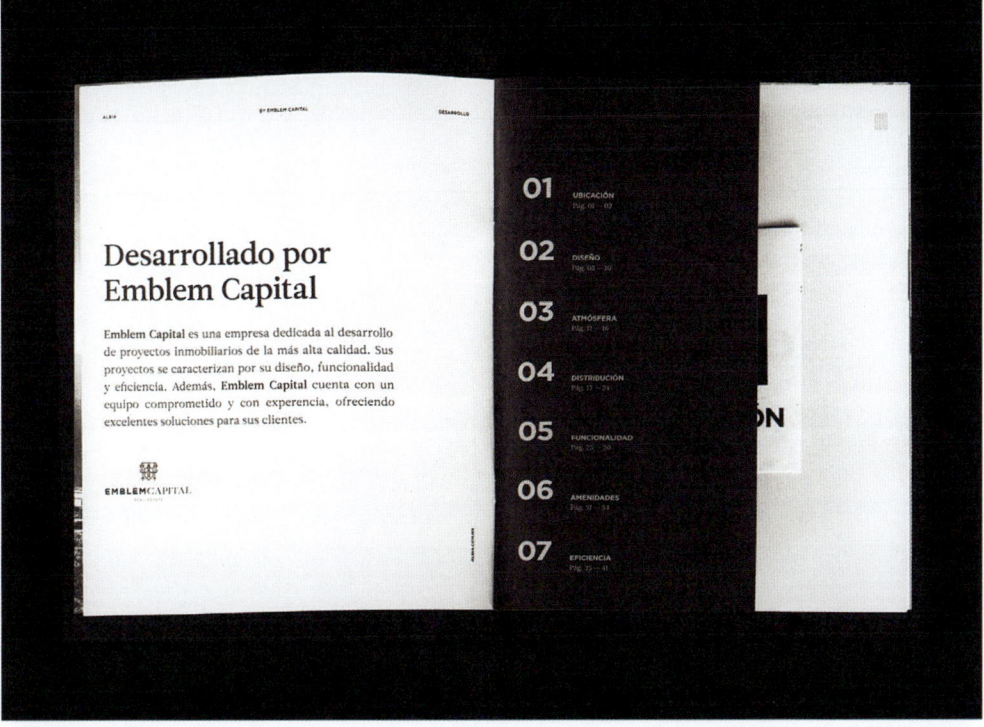

Un concepto inspirado en las necesidades de las nuevas empresas exitosas.

Desarrollado por Emblem Capital

Emblem Capital es una empresa dedicada al desarrollo de proyectos inmobiliarios de la más alta calidad. Sus proyectos se caracterizan por su diseño, funcionalidad y eficiencia. Además, **Emblem Capital** cuenta con un equipo comprometido y con experiencia, ofreciendo excelentes soluciones para sus clientes.

EMBLEMCAPITAL

WYNDHAM PLACE

CREDITS

Studio — **Phage Ltd.**
Client — **Studioloop**
Designer & Creative Director — **Natasha Zlobec,
Danny Brooks**

DESCRIPTION

Phage worked with property developers
Studioloop on a coffee-table book to launch their
latest luxury home to market. A Grade II listed
Georgian townhouse in the heart of London,
the property combines beautifully restored
original features with modern design and
home comforts, a juxtaposition of traditional
and contemporary that was a key differentiator
for potential buyers, and important to capture
in the design of the marketing materials.

We took one of the most striking features — a
beautifully restored fanlight — and developed it
into a visual motif for the book; foil-blocking it on
the front cover, de-bossing it on the title page,
and referencing it in the photography throughout,
helping to develop a narrative for the property,
and creating a running thread throughout the book.

The colour palette of delicate dove greys
and golds, and the use of textured stocks and
cover materials, were carefully sourced to
reference colours and finishes used throughout
property, while the typographic and image
style reflect a luxury lifestyle image, appealing
to the high net worth target audience.

16 WYNDHAM PLACE
LONDON W1

THE MASTER SUITE
—A LUXURY COCOON

ENJOY AN INDULGENT RETREAT IN THE MASTER
BEDROOM, DRESSING ROOM, AND ADJOINING
BATHROOM SUITE.

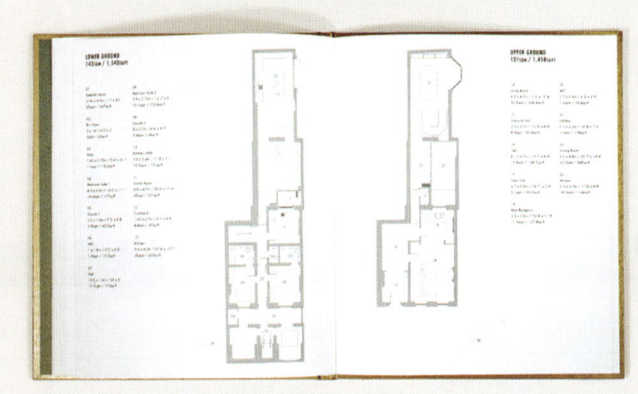

THE HOUSE
—PERFECTLY POISED

ENTRANCE AND RECEPTION
—FIRST IMPRESSIONS LAST

FROM THE MOMENT YOU SPOT THE HAND-
PAINTED HOUSE NUMBER, YOU ARE STRUCK BY
THE CONSIDERATION THAT HAS GONE INTO
EVERY LAST DETAIL OF THIS RENOVATION.

TITLE

42 WILLS
STREET

CREDITS

Studio — **StudioBrave**
Creative Director — **Tim Sutherland**
Designer — **Carlo Mussett**
Illustrator — **Carlo Mussett**

DESCRIPTION

Branding campaign for luxury residential property
development project in Glen Iris. Working closely
with John Demos Architects our core objective
was to create a striking identity that reflected
the architectural intent. The building's principles
were based on linear forms and layering.
Our creative expression mirrored these themes
through an experimental visual language
inspired by nature, rock and layers of the earth.

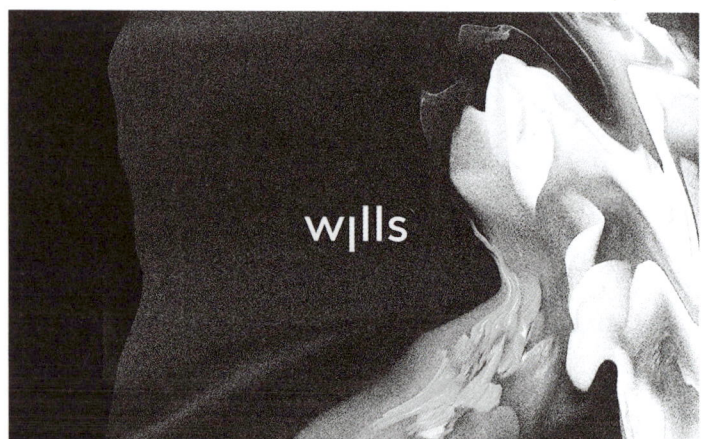

CHESTERFIELD HILL

CREDITS

Designer — **Campbell Hay**
Art Director — **Campbell Hay**

DESCRIPTION

The architects and property developers Vabel meticulously renovated a Georgian townhouse in London's Mayfair area to create a beautiful luxury home. They wanted a beautiful and sophisticated means of promoting the property.

Our answer was to create a narrative and art-direction that conveyed what it would be like to experience living at Chesterfield Hill and present it in a beautiful soft-bound publication.

Starting at the front, the brochure takes you on a curated walk through the house. While beginning at the end, presents a journey through Mayfair, providing further context and locality to the project. The two stories join seamlessly at the centre.

3 Chesterfield Hill
Mayfair, London
A project by Vabel

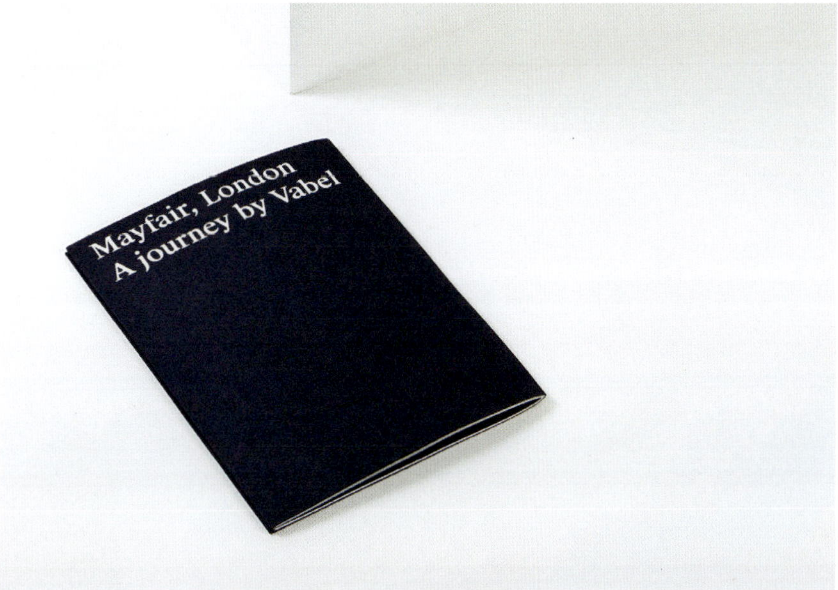

Mayfair, London
A journey by Vabel

THE HUNTLEY

CREDITS

Studio — **Born & Raised**
Creative Director — **Brad Stevens**
Designer — **Jordan Stokes, Jacqueline Waszkiewicz**
Writer — **Chris Laws**
Photographer — **Stephen Ward, Tim Jones**
Cinematographer — **Content Kitchen**
Motion Designer — **Content Kitchen**
Producer — **Chee Productions**

DESCRIPTION

Born & Raised developed name, brand identity
and international marketing campaign for a new
luxury residential resort and Greg Norman-
designed golf course, nestled just below the
Illawarra escarpment an hour south of Sydney.

The brand identity nods to the traditions of
golfing, balanced by a contemporary, luxurious
design language that celebrates the unique
position of this development — set up high and
surrounded by the beauty and colours of the
Australian bush, with dramatic views over
lake and ocean. The communications capture
the location's diversity and the promise of a
lifestyle that promises the best of everything.

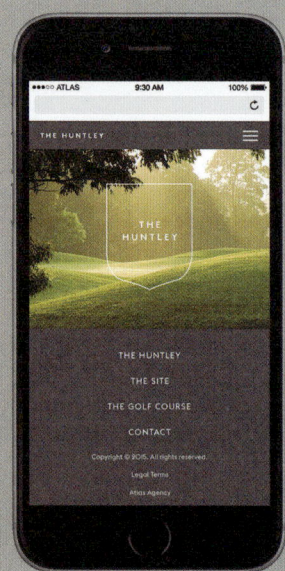

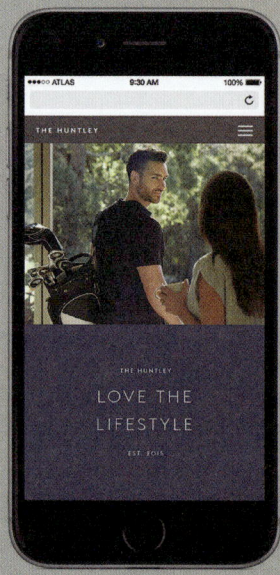

35 CLARENCE STREET

CREDITS

Studio — **THERE**
Designer — **Paul Tabouré,
Sophie Zetterburg, Jon Zhu**
Photographer — **THERE**
Client — **AMP CAPITAL GROUP**

DESCRIPTION

Rebranding and marketing a building with an idenitity crisis whilst creating a new sense of place. Commissioned by AMP Capital Group and partnering with CBRE Leasing, we created branded marketing materials for the leasing of over 7,385sqm tower space which included a number of contigious floors.

Through the creation of custom typography, carefully art-directed photography and crafted copywriting, a range of visually engaging lease marketing collateral was developed.

We focused on the 'openess' of the building in relationship to the city and surrounds, the contigious floorspace and the 270° views from the grade A tower.

where new possibilities open up
Endless possibilities and opportunities to maximise your 35 Clarence Street headquarters
Indicative Floorplans

ART RESIDENCE

CREDITS

Studio — **SmartHeart Branding Agency**
Creative Director — **Stas Okruh**
Art Director — **Yuriy Mihalchenko**
Strategist — **Ivan Kryukov**
Project Managers — **Artem Mitin, Anna Antonova**
Designers — **Konstantin Karpov, Sergey Kolesov,
Eugene Podlesnaya, Ilya Tumaikin**
Copywriters — **Maria Jelihovskaya,
Anna Antonova**
Coder — **Alexander Kuksin**
Web Developer — **Doctornet**

DESCRIPTION

Art Residence is a new design low-rise apartment
complex, which is located in residential area
in the center of Moscow just near by from
Tverskaya street. The client approached us
with the wish to create a modern expressive
identity and brand strategy to announce this
product appearance on Russian market.

The concept of the brand is an art as a part of
lifestyle and business, the unique opportunity
to invest in art and real estate simultaneously.
Visual identity continues the announced concept.
The logo reflects an architectural element of the
complex yellow squares, which are the part of
architecture building. It reflexes the art component
of the project, highlights the contemporary design
of residential area and symbolizes superior space
for comfort life. The art topic appears in the
names of the buildings: Art, Biennale, Concept,
Design, Etude, Form, Gallery, Harmony.

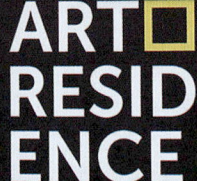

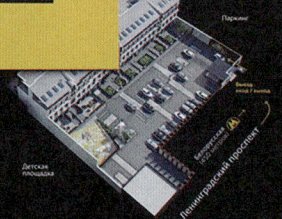

О ПРОЕКТЕ

ART□
RESID
ENCE

ДИЗАЙНЕРСКИЕ
АПАРТАМЕНТЫ
НА «БЕЛОРУССКОЙ»
5-я улица Ямского поля, дом 5

Art Residence — это принципиально новый дизайнерский жилой квартал премиум-класса, расположенный в центре Москвы, в двух шагах от Тверской. Art Residence — квартал для тех, кто дорого ценит стиль и функциональность. Это комплекс комфортных малоэтажных зданий, созданных по образу и подобию европейских городских особняков.

В таком пространстве как нигде можно ощутить все плюсы большого города, потому что здесь он не подавляет своими масштабами, а соразмерен и дружелюбен человеку. Атмосферу легкости создают отличающиеся безупречным вкусом пешеходные улочки и уникальные арт-объекты. Здесь действительно кажется, что вы в Европе.

LIVING IN THE ART OF MOS-COW

Искусство, с которым я живу

INVEST IN ART OF REAL ESTATE

Инвестиции в недвижимое искусство

РАЙОН

В СЕРДЦЕ ИСТОРИИ И ДЕЛОВОЙ ЖИЗНИ

Branding for Art Residence is characterized by a cool, contemporary style, high brand value, and exceptional combination of art & business attitudes. According to the main concept Art Residence is the first residential real estate in Moscow, which presents an art director as a main figure, who manages the cultural life of residents. He can clue residents in on going to the most preference events and receptions, give an advice on the concept of private party or provide an intelligent design solutions to resolve the most complicated interior design.

The advertising brochure and website artresidence.ru flesh out the main concept. The content is divided into 2 parts: Art & Residence — 2 completely different approaches and frames, which unites in one the main principle of brand concept. Art section releases the details of aesthetic aspect of the project (art & lifestyle), residence section describes in details the infrastructure and all the advantages of investments in real estate.

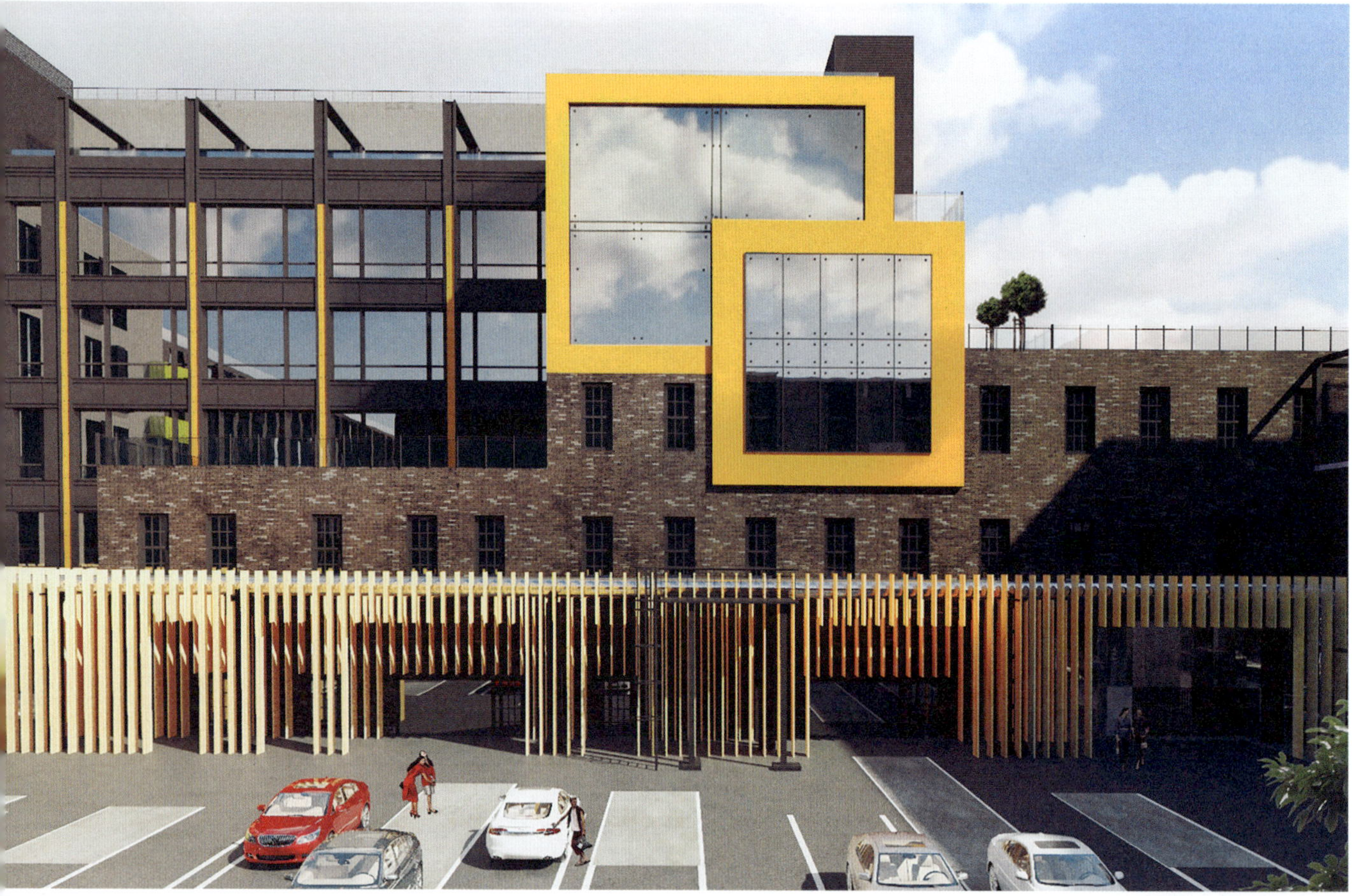

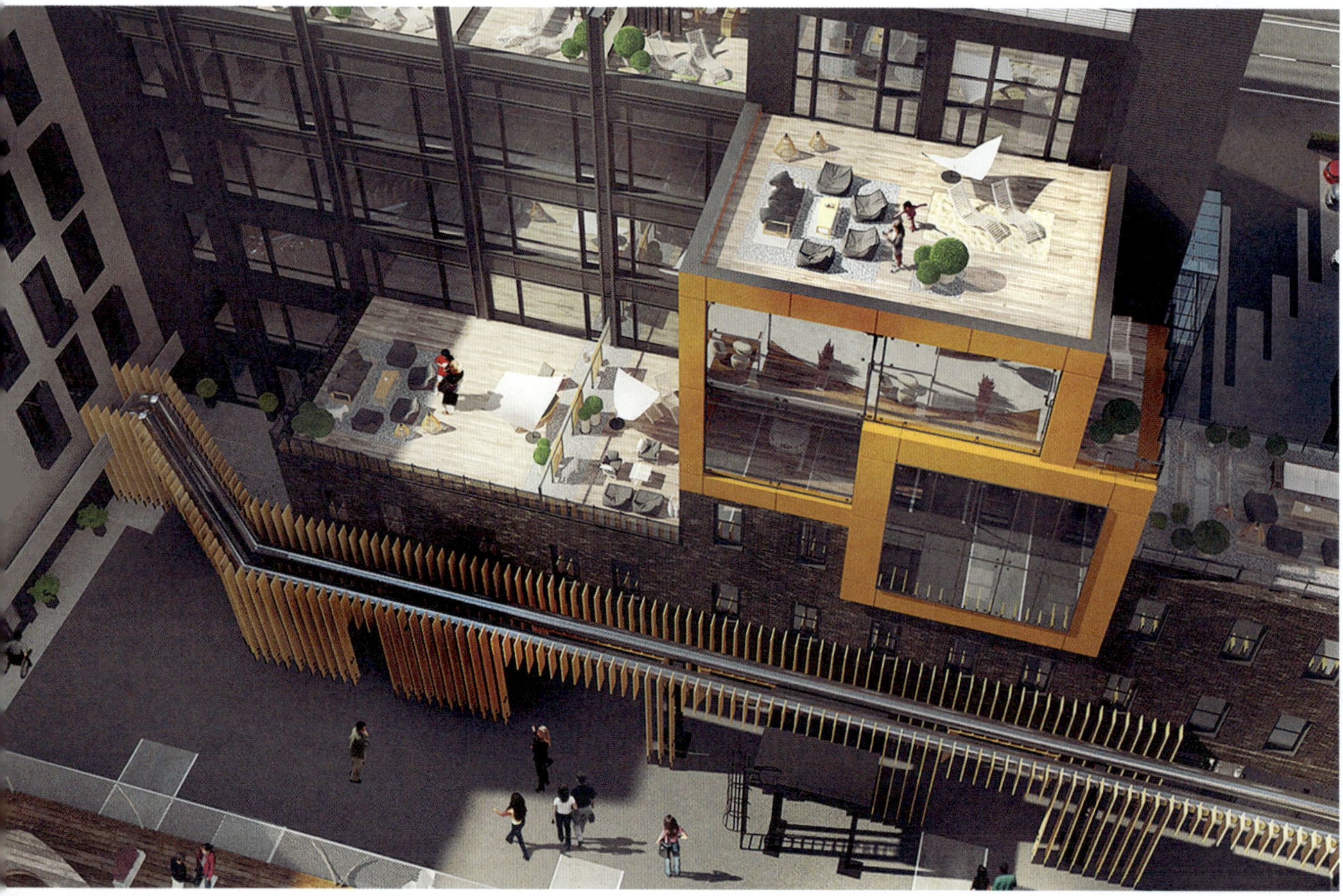

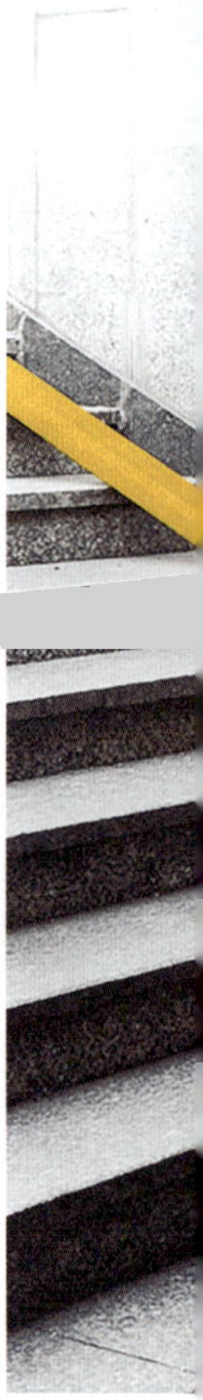

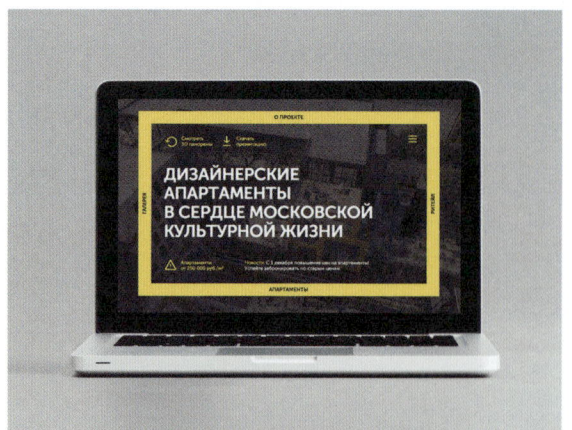

QUEENS PARK PLACE

CREDITS

Developer — **Londonewcastle**
(in partnership with London Borough of
Brent BY Development Limited)
Architect — **Ian Simpson Architects**
Interior Designer — **Tamzin Greenhill Design**
Interior Architect — **TG-Studio**
Art Director & Designer — **StudioSmall**
Photographer — **Mark Sanders**
Illustrator — **Klas Fahlén**

DESCRIPTION

Established in 1995, Londonewcastle is a multi-
million, London based, design-led residential
developer who over the past 20 years, has
established a name for working with the industry's
most visionary architects and designers.
We were commissioned to work on the brand
and communications for a new development of
luxury apartments in Queen's Park, London.
The project involved creating a complete
brand world that included naming creation,
identity, print, advertising and environment.

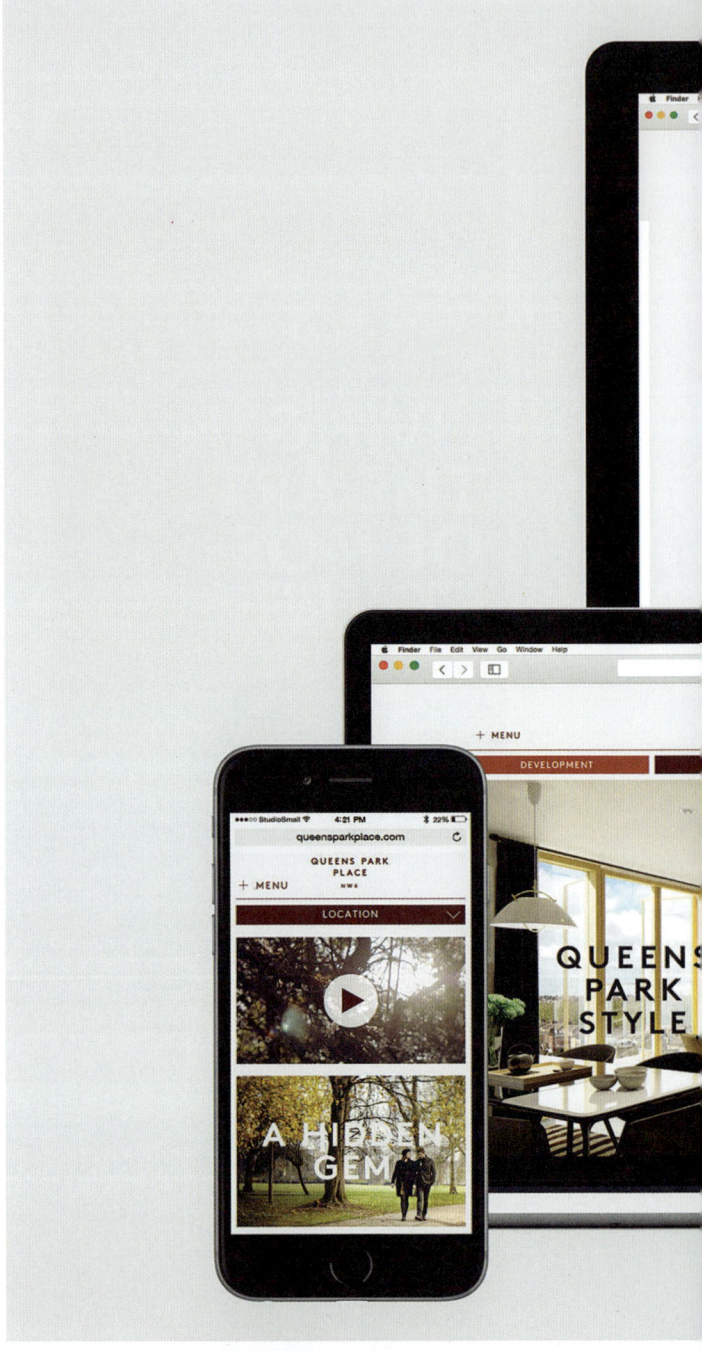

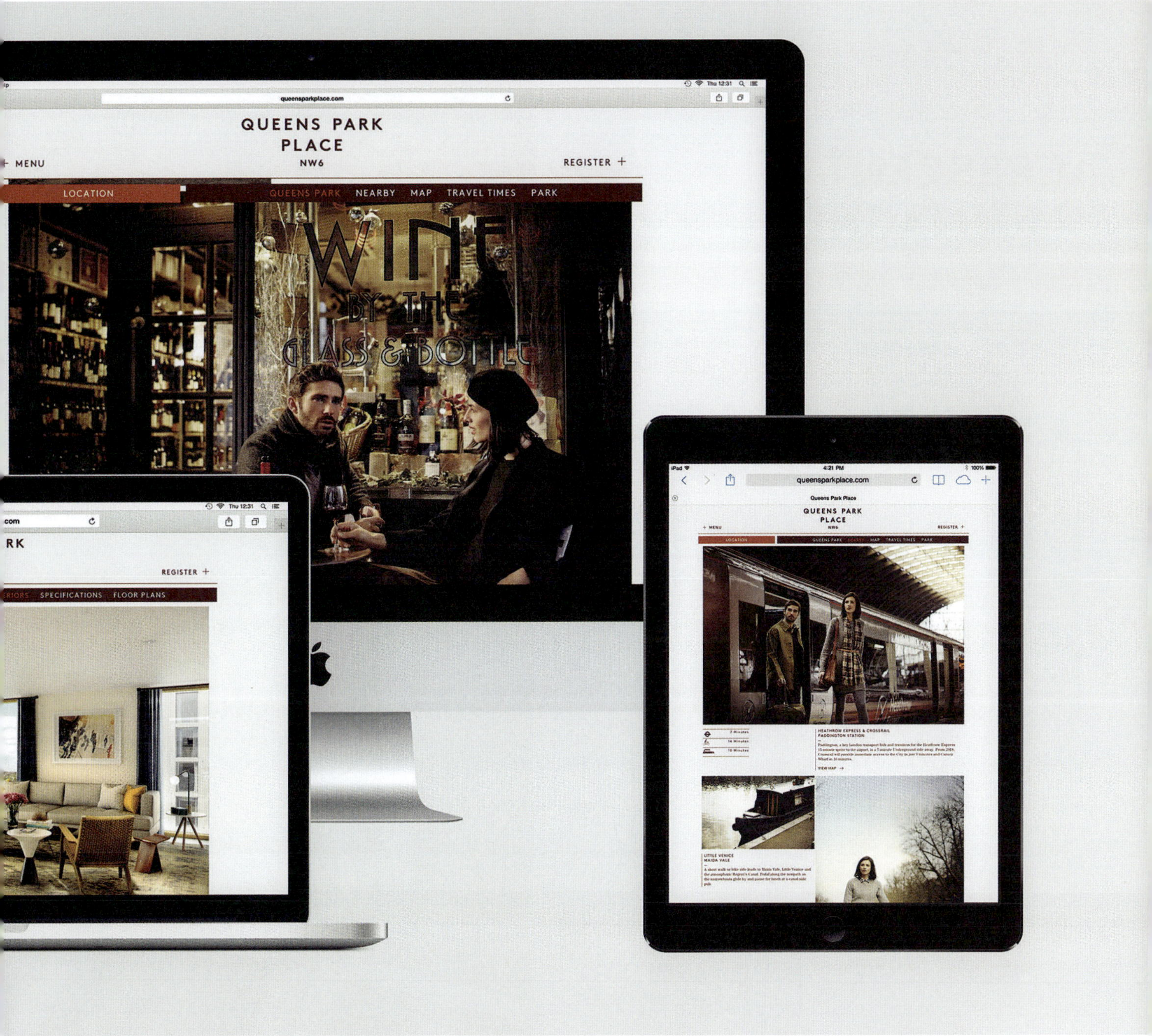

CRESCENT & MONTAGNE

CREDITS

Creative Director — **Christine Cook**
Copywriter — **Sophie Bordes**
Art Director — **Christine Cook, Victoria Di Valerio**
Designer — **Victoria Di Valerio**
Photographer — **Robert Viau**
Account Director — **Marie-Lou Rancourt**
Studio Director — **Annie-Claude Vachon**
Production Director — **Annie-Claude Vachon**

DESCRIPTION

Crescent and de la Montagne, two streets situated at the heart of the Golden Square Mile, are like no other in Montreal. Crofton Moore, a Canadian real-estate society, asked SGM to develop a strategy to promote leasing of the newly acquired properties owned by Lasalle. SGM developed an identity, allowing business that lease to become proud members of the C&M brand. It reflects the properties' notoriety and draws inspiration from the Victorian-style architecture to attract high-end brands. We developed a gold monogram using the letters C and M, representing the street names. The monogram is bold and regal while still remaining modern by coupling it with a sans serif typeface. The branding also includes a presentation folder, with descriptive inserts to be sent out to targeted businesses, and on site, elegant black and gold dressings for the buildings' window fronts.

CET ESPACE
N'ATTEND QUE VOUS

514 845-4500

CRESCENT & MONTAGN

CM

boutiques & bureaux

BOUTIQUES
DISPONIBLES

514 845-4500

LaSalle

GOLDEN SQUARE MILE

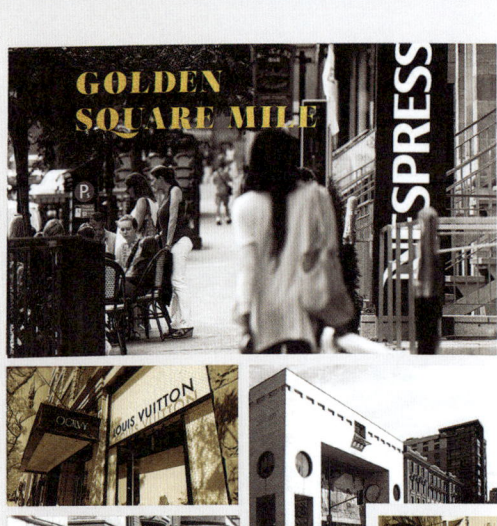

GOLDEN SQUARE MILE

14 ADRESSES DANS DEUX RUES PRESTIGIEUSES

Au cœur d'un quartier vivant, touristique et culturel, le charme des édifices historiques apporte une élégance authentique à l'image de votre marque.

14 ADDRESSES ON TWO PRESTIGIOUS STREETS

At the heart of a vibrant tourist and cultural district, where the stately elegance of historic buildings creates a fitting counterpart to your brand.

LOCATAIRES DE RENOM
BOUTIQUES DE LUXE
Clientèle raffinée

WELL-KNOWN TENANT
LUXURY BOUTIQUES
Refined clientele

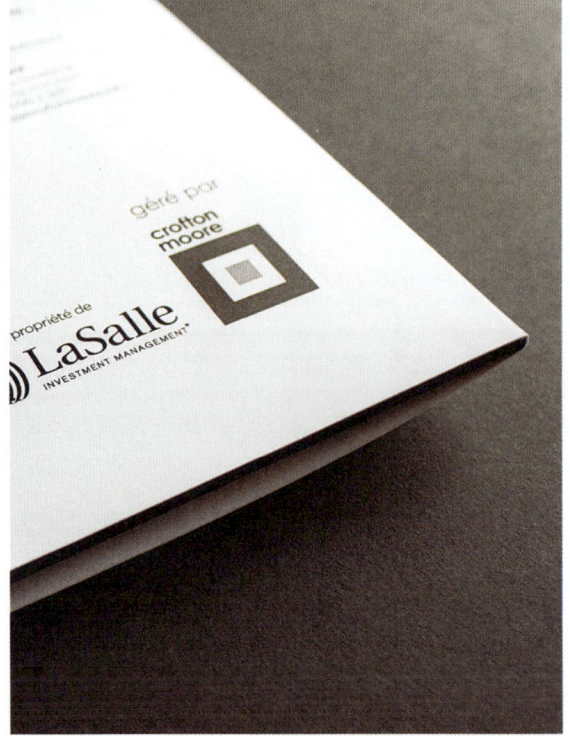

NORTEMÉRIDA

CREDITS

Studio — **FUTURA**
Editorial Photographer — **Fer Juaristi, Caroga**

DESCRIPTION

Nortemérida, is a residential development located in a privileged zone in Mérida, Mexico. We developed the project aiming to achieve a very clear visual communication. We created the slogan "La Ciudad que queremos," phrase that in Spanish has two different meanings: The City We Love, and The City We Want. Our intention was to play with this meanings, promoting that a better quality of life is possible.

The city of Mérida was our main inspiration, is a center of folklore with an enormous cultural wealth, and unparalleled gastronomy. It is a city that holds ancestral traditions and at the same time is inhabited by new generations of successful and well travelled individuals that have chosen Mérida as their home. All our graphic decisions for this project were based on the representation of this lifestyle.

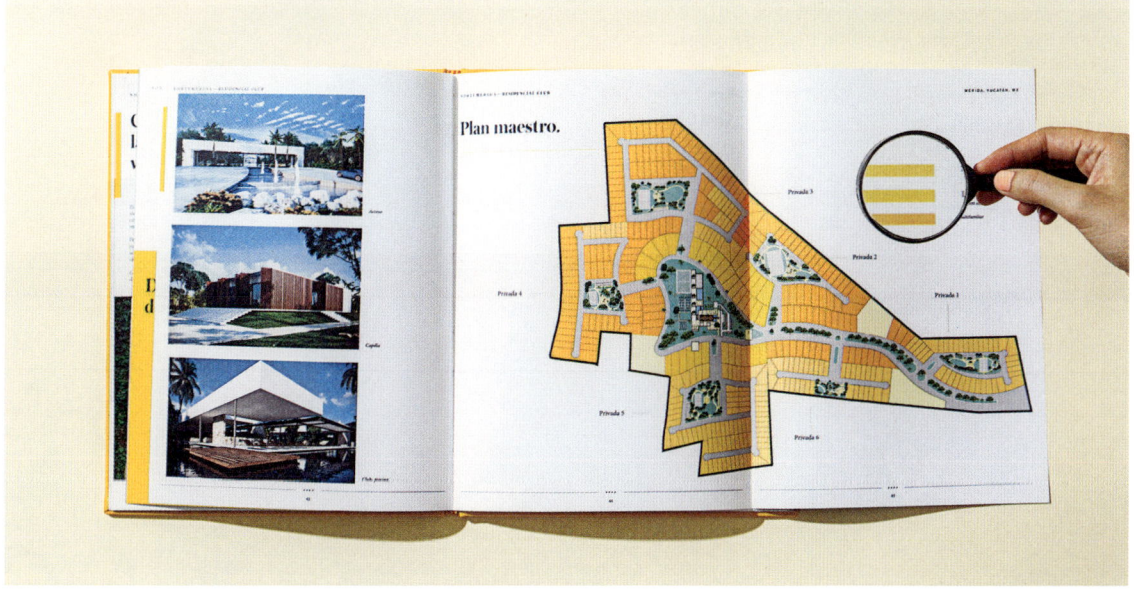

THE HUGO

CREDITS

Studio — **StudioBrave**
Creative Director — **Tim Sutherland**
Designer — **Jesse Mallon, Mike Nguyen**
Illustrator — **Andy Murray**
Photographer — **Sayher Heffernan**

DESCRIPTION

The Hugo is an iconic new multi residential development in emerging Footscray. We were engaged by Faymus to create a striking identity and marketing campaign that positioned the apartments in a very unique space. It was important that we captured the spirit behind the developers vision of the awakening of Footscray and its rich culture and diversity. We titled the project Footscray Reimagined.

We commissioned local artist Andy Murray (Gatsby) to illustrate our visions of a vibrant Footscray lifestyle. The collaboration added great depth and authenticity to the project. We also engaged photographer Sayher Heffernan to capture a rare and beautifully honest side of the West.

THE
HUGO

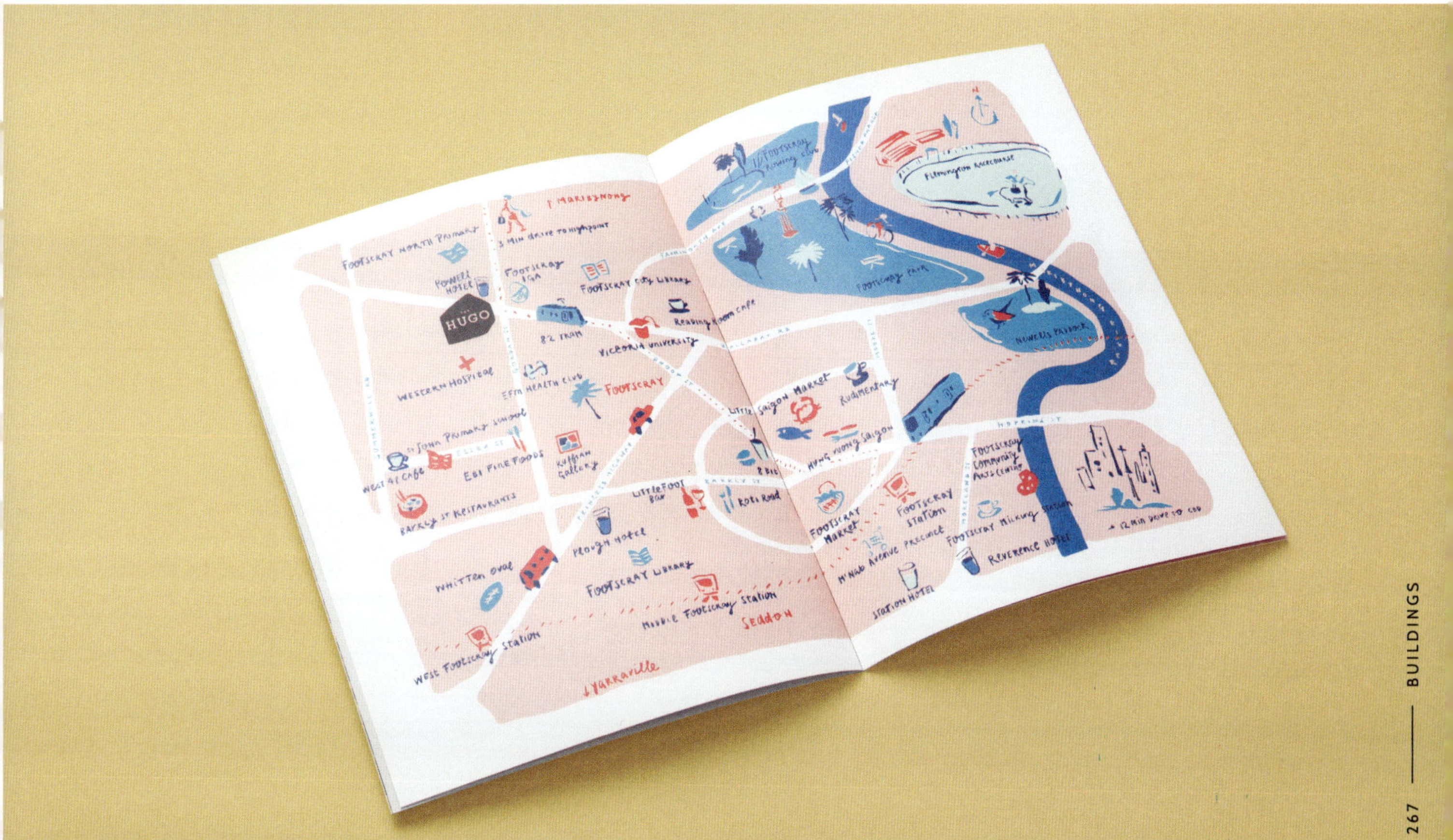
HUGO
FOOTSCRAY
Little Saigon Market
Footscray Park
Flemington Racecourse
Victoria University
82 Tram
Western Hospital
Footscray Library
Footscray Station
Footscray Market
Kori Road
Reverence Hotel
Station Hotel
Whitten Oval
West Footscray Station
Middle Footscray Station
Seddon
Yarraville

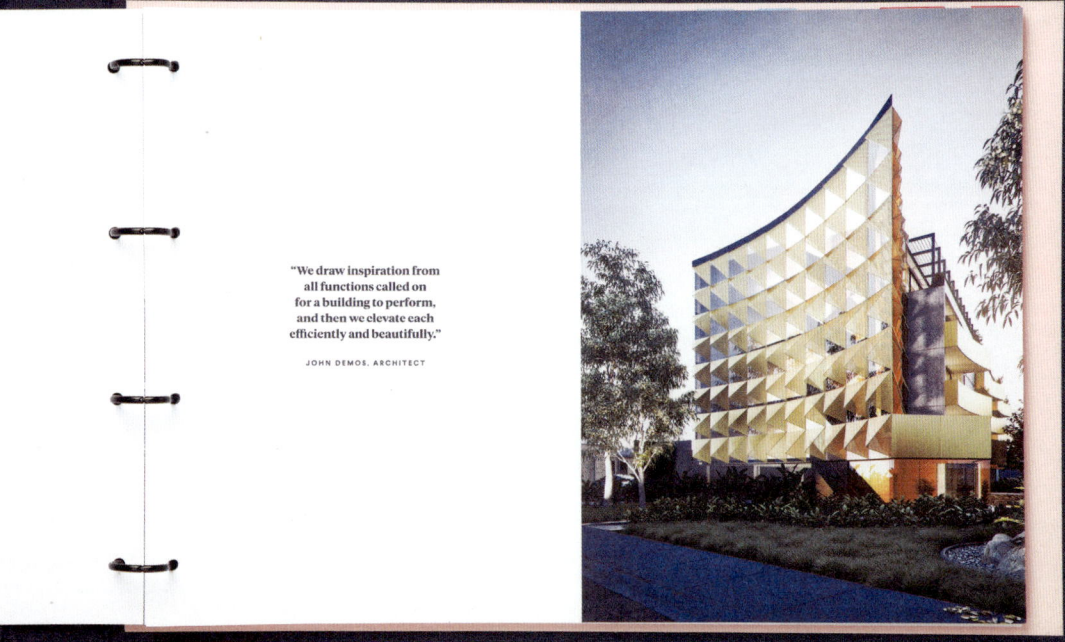

"We draw inspiration from all functions called on for a building to perform, and then we elevate each efficiently and beautifully."

JOHN DEMOS, ARCHITECT

Famous food markets
and its burgeoning
café scene.

A COMPELLING DESTINATION

TITLE

DE NODE

CREDITS

Studio — **THERE**
Designer — **Paul Tabouré, Jon Zhu**
Photographer — **THERE**
Client — **Haralambis Property**

DESCRIPTION

Surry Hills has long been known as a vibrant and emerging district of Sydney – after all that's where our creative studio is based! So we were excited when commissioned by Haralambis Developments to create the branding and sales marketing collateral for this boutique retail/residential mixed use development, designed by renown architects, Candalepas.

The scope included building identity, sales brochure, website, flyers and sales suite. The sucess of the marketing campaign saw all sales completed ahead of schedule.

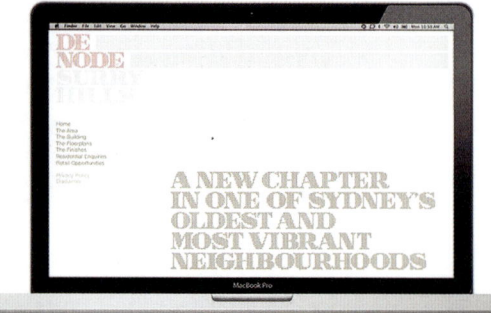

CHALET JANLUKE

CREDITS

Studio — **Phage Ltd.**
Designer & Creative Director — **Natasha Zlobec, Danny Brooks**

DESCRIPTION

Chalet Janluke is a luxury ski chalet in Switzerland. The architects, Richard Mitzman, approached Phage to design a brand and website to attract potential customers and to use on-site across the chalet interiors.

Combining traditional and modern materials, building techniques, and styles, Chalet Janluke remodels elements found in traditional Swiss chalets in a contemporary way. This blending of old and new was carried through to the brand, with a typographic style and colour palette evocative of 1950s Swiss tourist board posters combined with a fluid, interchangeable snowflake pattern that creates a fresh, modern feel with its graphic simplicity.

The pattern, which forms the foundation of the whole brand system, is composed of individual snowflake designs that interlock and build up around the logo in a potentially infinite number of fractal-type arrangements, reminiscent of frost patterns on glass. These compositions were applied to frosted glass, exterior signage, and wooden window shutters, as well as to the marketing materials.

Image 2/4
Chalet Janluke looks over La Tzoumaz
La Tzoumaz

Designed by award-winning architects Richard Mitzman Architects and situated at an altitude of over 1550m overlooking the beautiful Rhone Valley, Chalet Janluke is a newly built luxurious chalet in La Tzoumaz, part of the fabulous Verbier Four Valleys ski system.

LEOPOLDO MARECHAL

CREDITS

Studio — **The Negra**
Project Director — **The Negra**
Creative Director — **The Negra**

DESCRIPTION

The real estate project is placed at Leopoldo Marechal Street, its name refers to a famous Argentine writer. Marechal was a poet, playwright, novelist and essayist, someone who lives in a constant creativity process of bringing thoughts to paper.

For the logo, we investigate and decided to join this kind of life style whit dream's universe. Dreams that can be realize just with a signature (and an amount of money to support it). The dream and the effort of getting a home ownership.

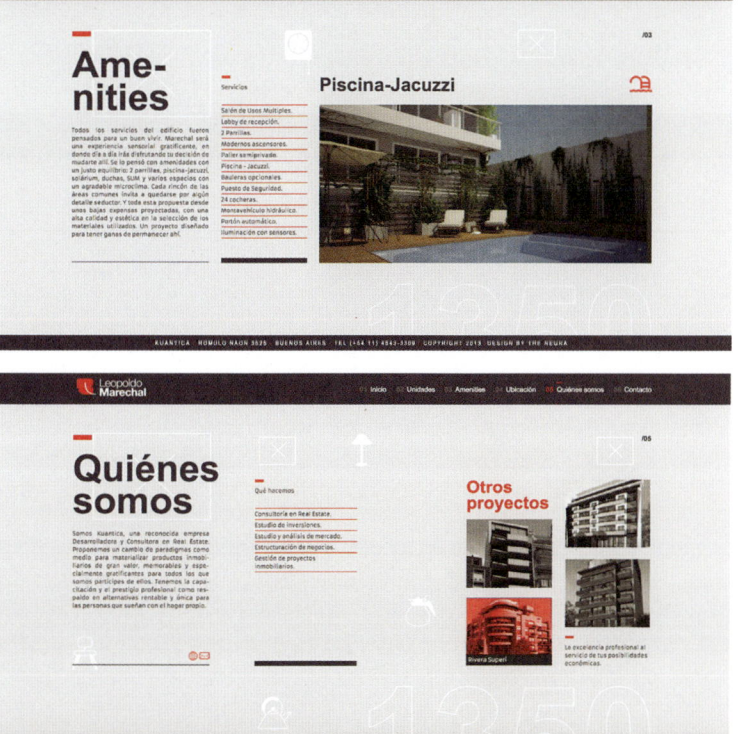

ALMOND APARTMENTS

CREDITS

Studio — **Hooga Creative**
Creative Director — **Dmytro Yarynych**
Art Director — **Denys Kuzmenko**
Designer — **Kyrylol Shvedov, Tetiana Kobryn**

DESCRIPTION

Almond apartments is a residential complex in Kyiv, Ukraine. Situated in underdeveloped neighbourhood, it is, however, a great investment — 10 minutes walking distance to the city underground station, a brand new infrastructure, and a lot of other goodies (well, we don't sell any). Last, but not least — the lowest price rate in Kyiv, we mean, wow. We developed a brand name, communication and basic philosophy to help our client show all the benefits of buying apartments in their newest development.

Main visual motive — is everything around almond, it's flowers, smell, texture and colour. It is all about scent of spring on the street, cosy aroma candles in your living room, tasty holiday cake or Sunday morning croissant — all of it was kept in our heads in the process of design, and we are pretty sure that potential residents will feel that too.

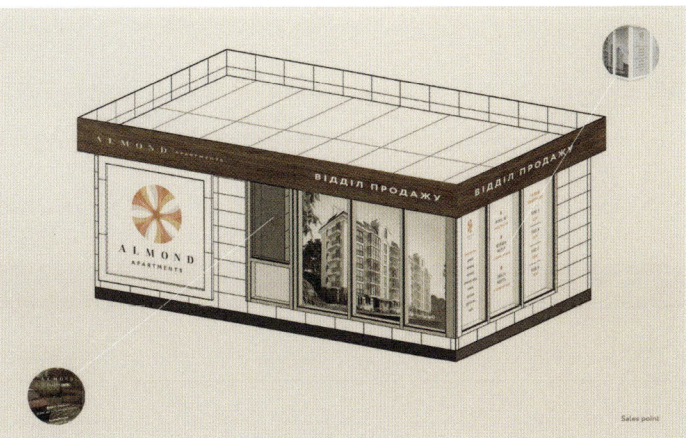

LUNADA

CREDITS

Studio — **FUTURA**

DESCRIPTION

Lunada is a brand-new concept of apartments in Playa del Carmen. Our main objective was to communicate that the Mexican Caribbean is much more than white sand and a transparent sea. The Caribbean is vegetation, music, flavours and it offers a lifestyle that is impossible to understand unless you are there.

In this occasion, apart from the branding, editorial, collateral design and website, we worked on the interior design of the sample apartment. This allowed us to expand the branding until its limits, turning it into materials and construction finishes, distribution, lightning, textiles, furniture and decorative objects.

Developing the interior design based on the branding, makes easier the selling, because once the potential clients are inside the apartment, they can truly experience the whole concept.

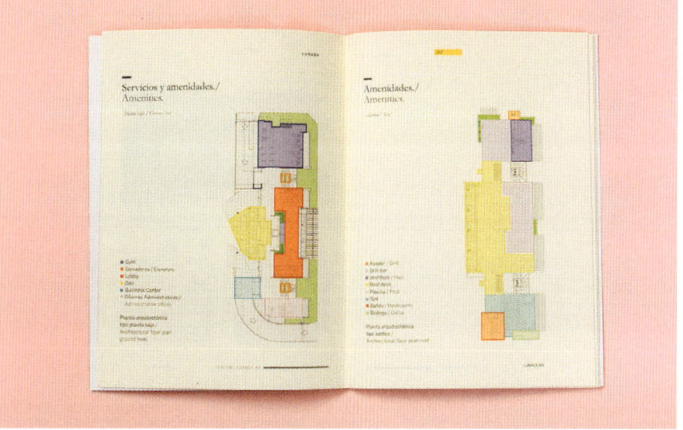

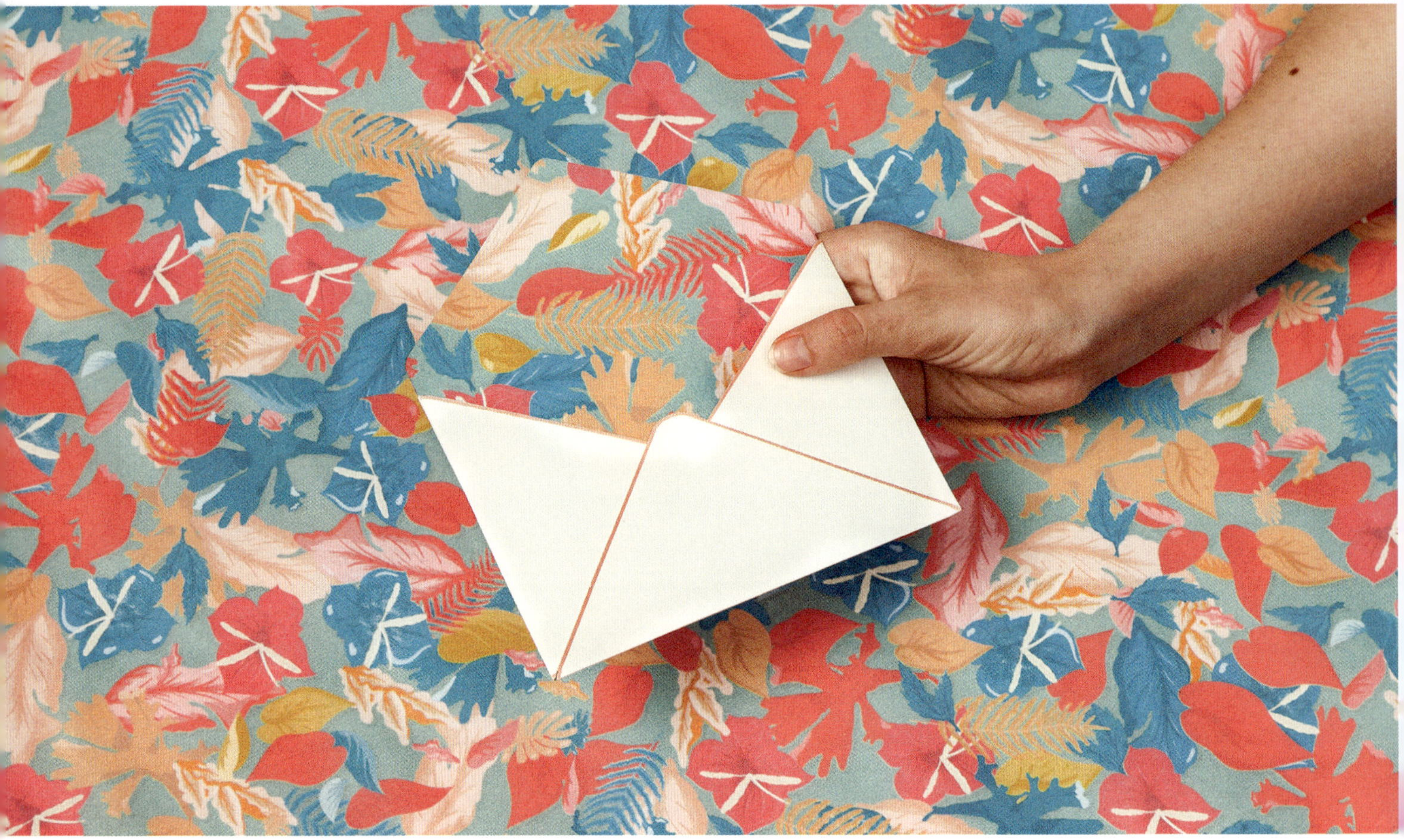

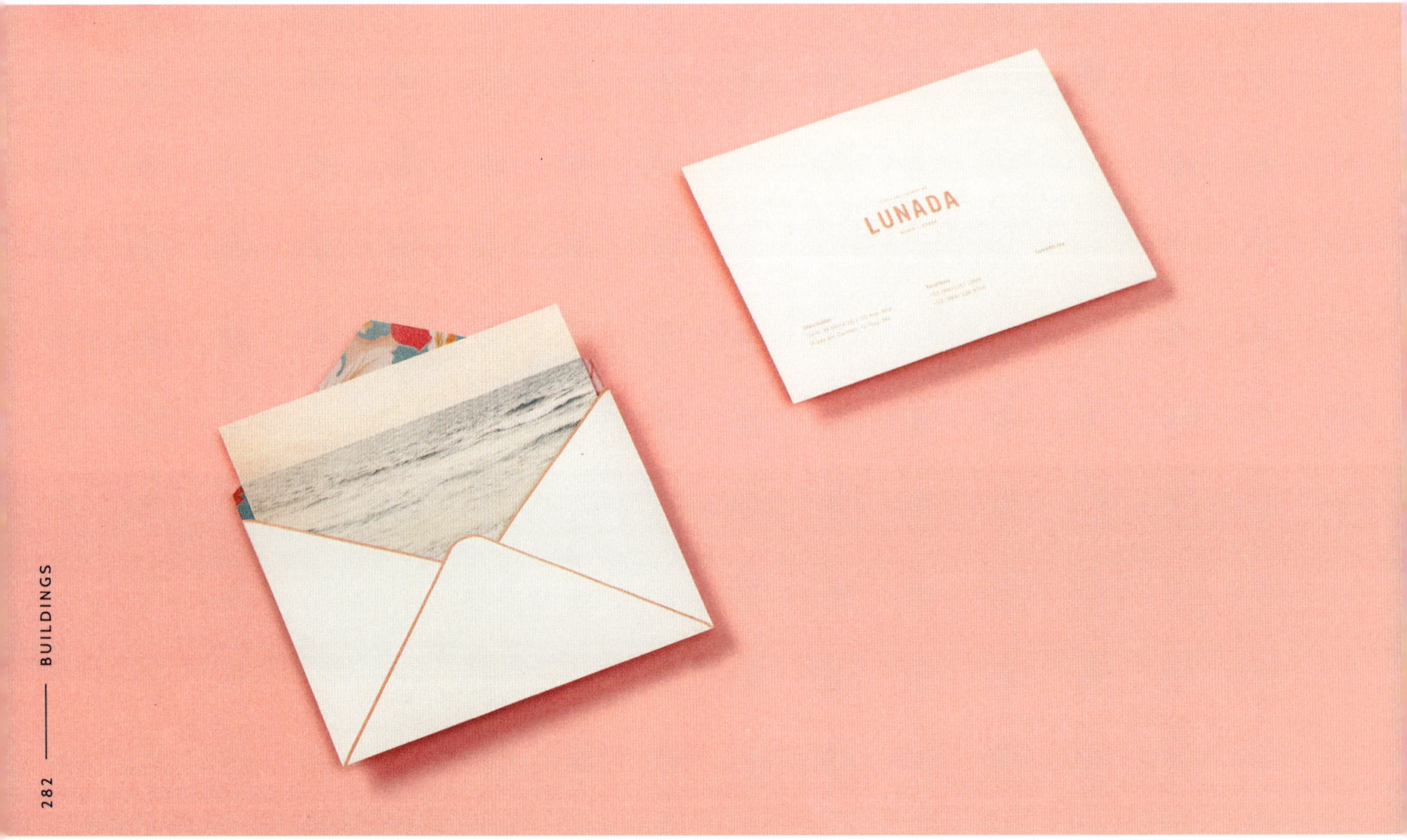
LUNADA

TITLE

ERA

CREDITS

Studio — **THERE**
Designer — **Paul Tabouré, Simon Hancock**
Photographer — **THERE**
Client — **Hayson Group**

DESCRIPTION

This mixed use commercial development by Hayson Group, consisted of retail shops, supermarket and 45 comtemporary office studios. It was the first new building development in the Kings Cross area for nearly 20yrs.

Stanisic Architects designed a building for the future with a very high environmental rating. We designed a brand image and marketing campaign that reflected this futuristic vision showcasing the buildings key aspects and appealing to a forward thinking audience.

SEE THE
FUTURE

ERA
ERA ON SPRINGFIELD

— SEE
THE
FUTURE

— UNFOLD

FWD »

>>
FLOOR PLAN
Office Type B1/B

OFFICE SUITES
105, 205, 305,
305, 305, 405,
405, 505, 505
AREA
80 sqm

ERA
ERA ON SPRINGFIELD

>> FFWD TO THE FUTURE

© 1993 FORD
IN LEASE NEW PRG
SPRINGFIELD STREET
TOWN APARTMENT
NSW 2025

FLOOR PLAN
Office Type A2

AREA
51 sqm

FFWD
TO
T
FUTU

A NEW ER
A NEW F

DEPOSIT A
NEXT EA

ERA
ERA ON SPRINGFIELD

Each of the 44 office suites embrace the social and
natural environment in ways that will make them a joy
to work in and a wise financial choice to own.
From the outside, the streamlined lines and
graphite grey, silver and blood red exterior presents a
bold face to the world.
Everything has been considered. Changing
needs. Opportunities for interaction. Access
demographics. The shifting nature of work. Access
to transport. Technological infrastructure. Physical
beauty. Life.

ERA. The face of the future.

EMBRACE THE FUTURE TODAY
IN A BUILDING DESIGNED
R TOMORROW

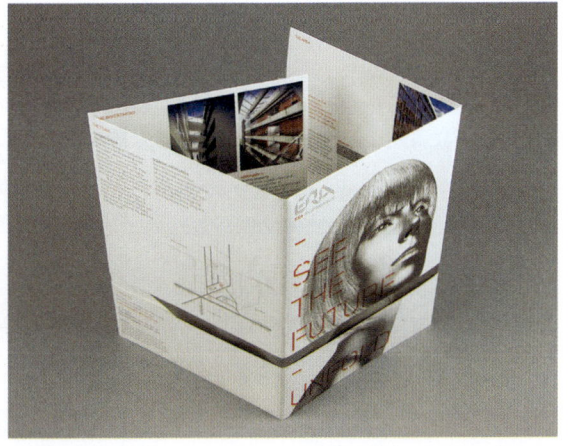

ONE COMMERCIAL STREET

CREDITS

Studio — **Journal**
Designer — **Paul Spencer, John-Paul Warner**
Photographer — **Kerry Harrison**

DESCRIPTION

One Commercial Street is a mixed use development nestled in London's diverse eastern city fringe. It's close to the traditional finance and legal districts, but also on the edge of the avant-garde, creative area of Shoreditch. This created a quandary for the developer, Redrow London — who should they target?

Journal were tasked with the creation of a brand and campaign to market the commercial office space within the building. The main challenge was to hit the right note and develop a personality which could communicate effectively to a wide and varied audience. The solution played on the number in the building's name — using differently styled number 1's to represent the potential for a range of business sectors under one roof, as well as the different possibilities for the space itself. This was supported with the tag-line "one for all." The numbers were used throughout the marketing of One Commercial Street — from the property brochure through to interior signage and 3D installations.

One for all nearby

When it comes to location, every business has a different list of boxes they need ticking. Lying on the edge of both the City and the City Fringe, One Commercial Street will fill them all.

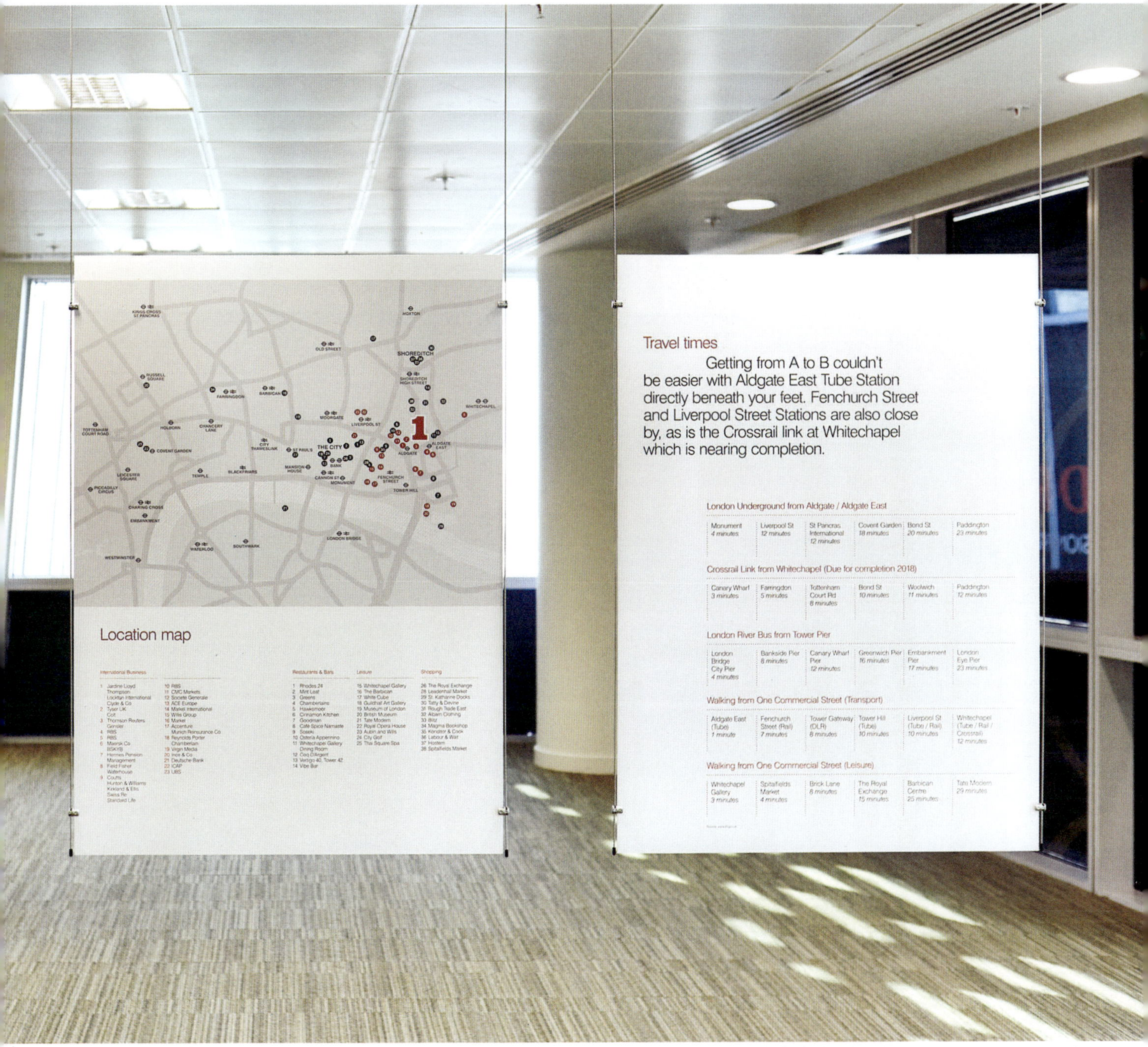

Location map

International Business
1 Jardine Lloyd Thompson
Lockton International
Clyde & Co
Tyser UK
Coil
3 Thomson Reuters
Gensler
4 RBS
5 RBS
6 Maersk Oil
BSKYB
7 Hermes Pension Management
8 Field Fisher Waterhouse
9 Coutts
Hunton & Williams
Kirkland & Ellis
Swiss Re
Standard Life

10 RBS
11 CMC Markets
12 Societe Generale
13 ACE Europe
14 Markel International
15 Willis Group
16 Markel
17 Accenture
Munich Reinsurance Co
18 Reynolds Porter Chamberlain
19 Virgin Media
20 Inox & Co
21 Deutsche Bank
22 ICAP
23 UBS

Restaurants & Bars
1 Rhodes 24
2 Mint Leaf
3 Greens
4 Chamberlains
5 Hawksmoor
6 Cinnaman Kitchen
7 Goodman
8 Café Space Namaste
9 Sasaku
10 Osteria Appennino
11 Whitechapel Gallery Dining Room
12 Coq D'Argent
13 Vertigo 42, Tower 42
14 Vibe Bar

Leisure
16 Whitechapel Gallery
17 The Barbican
18 White Cube
19 Guildhall Art Gallery
20 Museum of London
21 British Museum
22 Tate Modern
23 Royal Opera House
24 Aubin and Wills
25 City Golf
25 Thai Square Spa

Shopping
26 The Royal Exchange
27 Leadenhall Market
28 The Royal Exchange
29 St. Katharine Docks
30 Tatty & Devine
31 Rough Trade East
32 Albam Clothing
33 Blitz
34 Magma Bookshop
35 Kinston & Cook
36 Labour & Wait
37 Hostem
38 Spitalfields Market

Travel times

Getting from A to B couldn't be easier with Aldgate East Tube Station directly beneath your feet. Fenchurch Street and Liverpool Street Stations are also close by, as is the Crossrail link at Whitechapel which is nearing completion.

London Underground from Aldgate / Aldgate East

Monument	Liverpool St	St Pancras International	Covent Garden	Bond St	Paddington
4 minutes	12 minutes	12 minutes	18 minutes	20 minutes	23 minutes

Crossrail Link from Whitechapel (Due for completion 2018)

Canary Wharf	Farringdon	Tottenham Court Rd	Bond St	Woolwich	Paddington
3 minutes	5 minutes	8 minutes	10 minutes	11 minutes	12 minutes

London River Bus from Tower Pier

London Bridge City Pier	Bankside Pier	Canary Wharf Pier	Greenwich Pier	Embankment Pier	London Eye Pier
	8 minutes	12 minutes	16 minutes	17 minutes	23 minutes

Walking from One Commercial Street (Transport)

Aldgate East (Tube)	Fenchurch Street (Rail)	Tower Gateway (DLR)	Tower Hill (Tube)	Liverpool St (Tube / Rail)	Whitechapel (Tube / Rail / Crossrail)
1 minute	7 minutes	8 minutes	10 minutes	10 minutes	12 minutes

Walking from One Commercial Street (Leisure)

Whitechapel Gallery	Spitalfields Market	Brick Lane	The Royal Exchange	Barbican Centre	Tate Modern
3 minutes	4 minutes	8 minutes	15 minutes	25 minutes	29 minutes

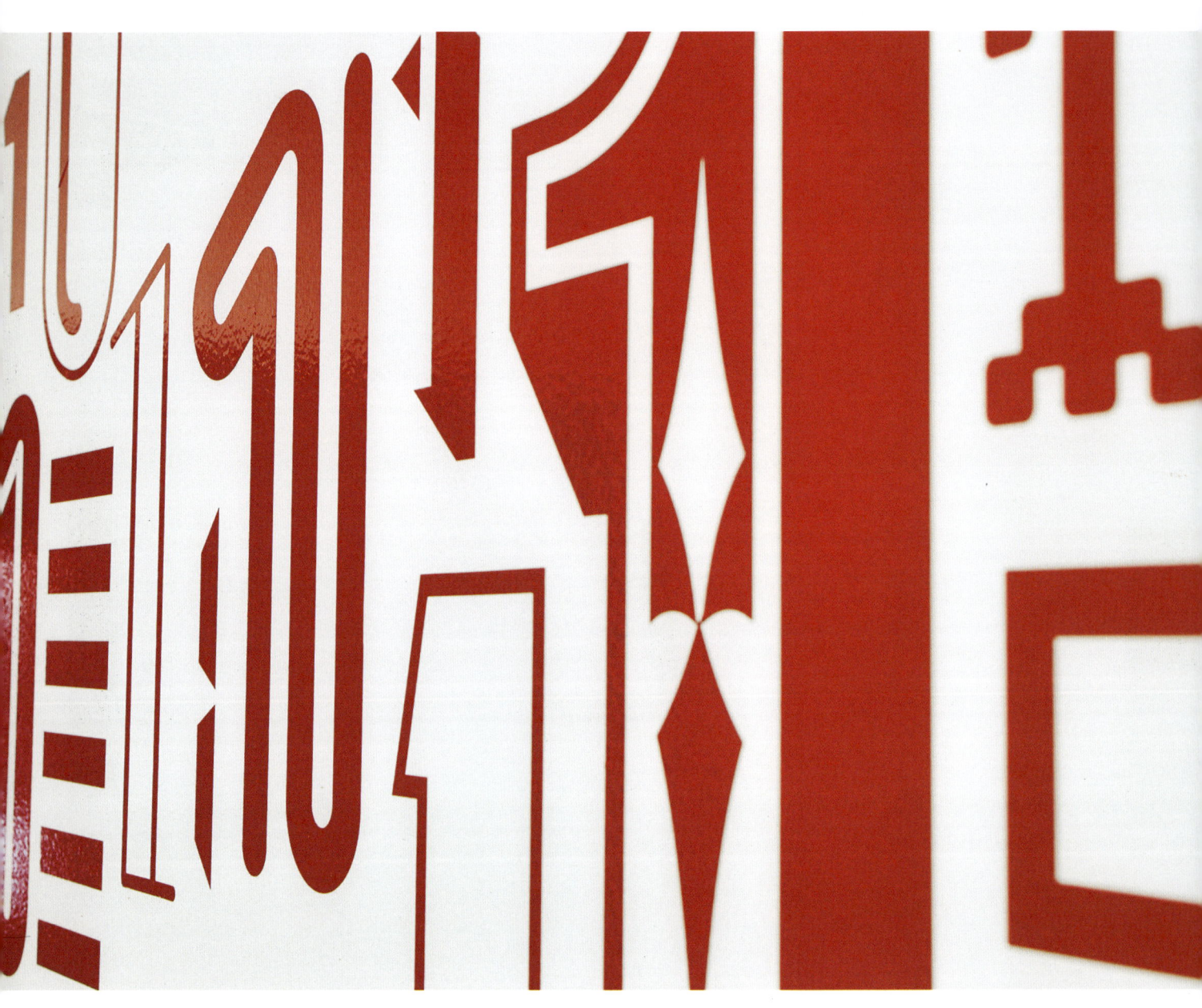

One for all your needs

Businesses come in all shapes and sizes. That's why the office space at One Commercial Street is adaptable to your business requirements.

One for Digital

Industry will find new standards around Commercial Street's Innovation Technology Development and is able to meet the needs of all shapes and sizes, the exacting expectations that is needed in every field, as working on what you need from your business. Where next shall Sovereign needs problem? How about a games room? You've got it.

338 PITT STREET

CREDITS

Studio — **THERE**
Designer — **Paul Tabouré, Teresa Luckman,
Sophie Zetterburg, Jon Zhu**
Photographer — **THERE**
Client — **AMP CAPITAL GROUP**

DESCRIPTION

We were commissioned by AMP Capital
Group to partner with CBRE Leasing, and help
reposition 338 Pitt Street – a commercial office
tower in need of an identity refresh. We focused
on the culturally diverse and buzzing heart
of the local vibrant district and its centrality
to well, everything! Touchpoints included
building signage, leasing boards, flyers,
sales brochure and tenant rep templates.

EVERYTHING
ALIGNS AT
338 PITT STREET

THE RICH AND VARIED MIX OF URBAN
AND PARKLAND ENVIRONMENTS WILL
IMPROVE THE WORK/LIFE
BALANCE OF STAFF.

BUILD BETTER
BALANCE

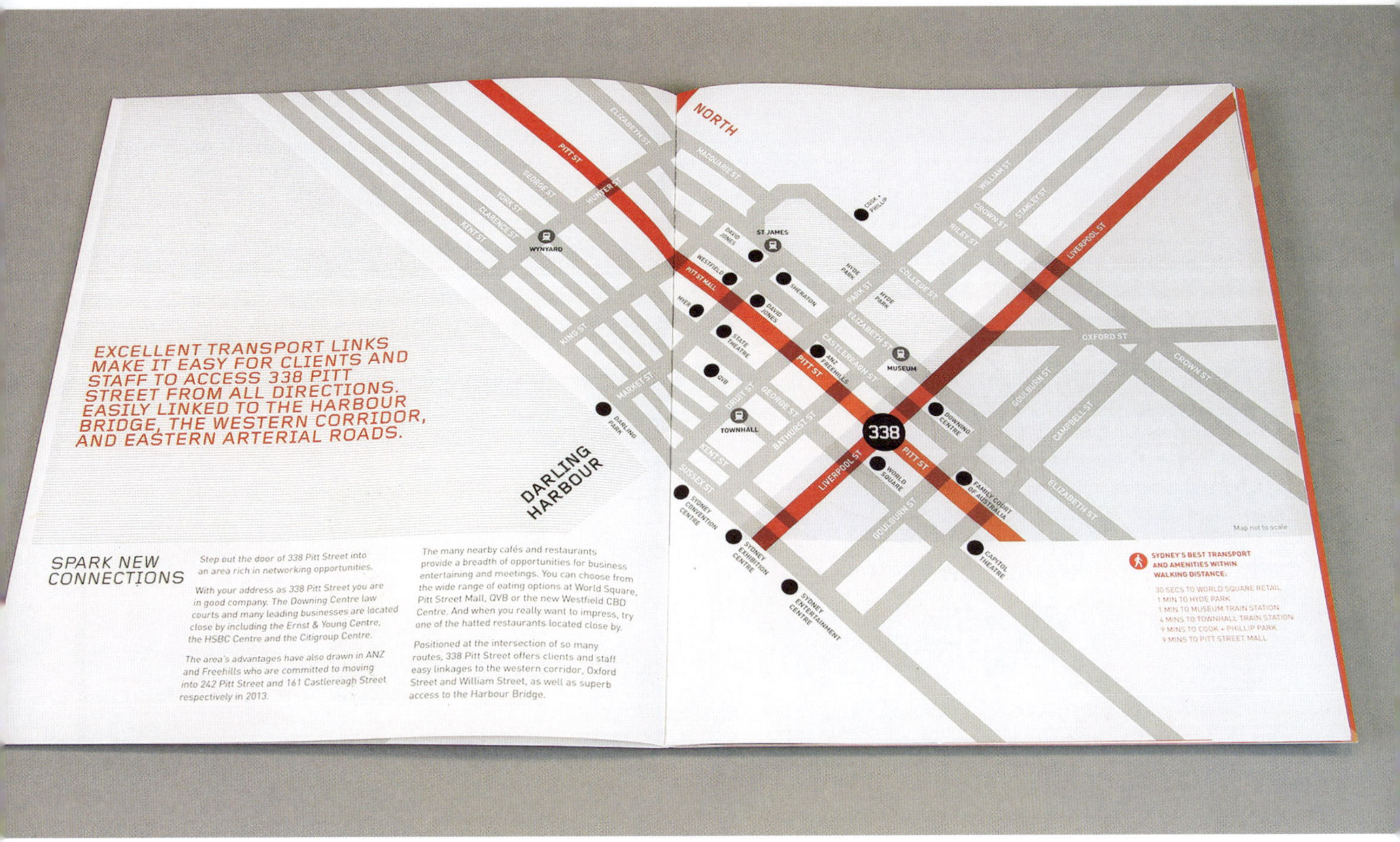

NORTH

EXCELLENT TRANSPORT LINKS
MAKE IT EASY FOR CLIENTS AND
STAFF TO ACCESS 338 PITT
STREET FROM ALL DIRECTIONS.
EASILY LINKED TO THE HARBOUR
BRIDGE, THE WESTERN CORRIDOR,
AND EASTERN ARTERIAL ROADS.

DARLING
HARBOUR

338

SPARK NEW CONNECTIONS

Step out the door of 338 Pitt Street into an area rich in networking opportunities.

With your address as 338 Pitt Street you are in good company. The Downing Centre law courts and many leading businesses are located close by including the Ernst & Young Centre, the HSBC Centre and the Citigroup Centre.

The area's advantages have also drawn in ANZ and Freehills who are committed to moving into 242 Pitt Street and 161 Castlereagh Street respectively in 2013.

The many nearby cafés and restaurants provide a breadth of opportunities for business entertaining and meetings. You can choose from the wide range of eating options at World Square, Pitt Street Mall, QVB or the new Westfield CBD Centre. And when you really want to impress, try one of the hatted restaurants located close by.

Positioned at the intersection of so many routes, 338 Pitt Street offers clients and staff easy linkages to the western corridor, Oxford Street and William Street, as well as superb access to the Harbour Bridge.

Map not to scale

**SYDNEY'S BEST TRANSPORT
AND AMENITIES WITHIN
WALKING DISTANCE:**

30 SECS TO WORLD SQUARE RETAIL
1 MIN TO HYDE PARK
1 MIN TO MUSEUM TRAIN STATION
4 MINS TO TOWNHALL TRAIN STATION
9 MINS TO COOK + PHILLIP PARK
9 MINS TO PITT STREET MALL

GOOD ARCHITECTURE
WORKS FROM TWO
ANGLES: FAR AWAY
AND UP CLOSE...

EVERYTHING
ALIGNS AT
338 PITT STREET

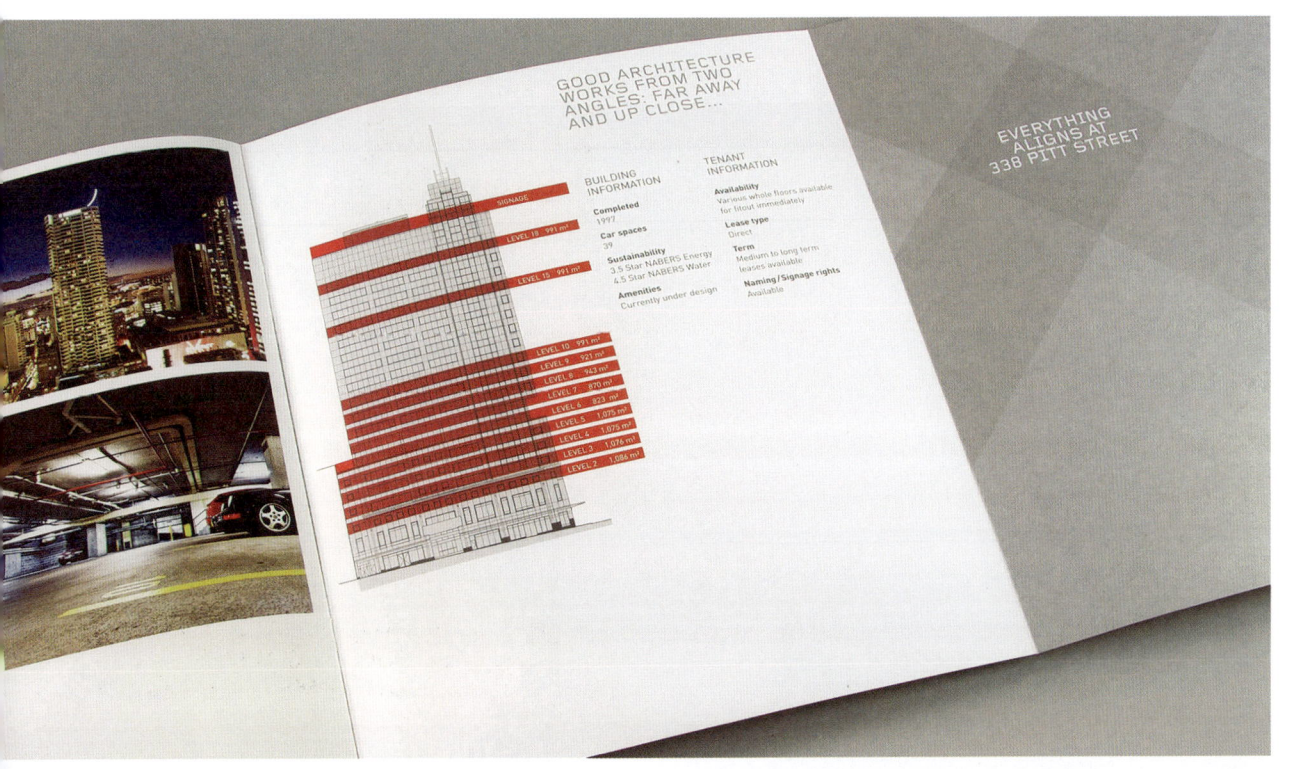

BUILDING INFORMATION

Completed
1997

Car spaces
39

Sustainability
3.5 Star NABERS Energy
4.5 Star NABERS Water

Amenities
Currently under design

TENANT INFORMATION

Availability
Various whole floors available
for fitout immediately

Lease type
Direct

Term
Medium to long term
leases available

Naming / Signage rights
Available

SIGNAGE
LEVEL 16   991 m²
LEVEL 15   991 m²
LEVEL 10   991 m²
LEVEL 9   921 m²
LEVEL 8   943 m²
LEVEL 7   870 m²
LEVEL 6   823 m²
LEVEL 5   1,075 m²
LEVEL 4   1,075 m²
LEVEL 3   1,076 m²
LEVEL 2   1,086 m²

AM FLEISCHMARKT

CREDITS

Studio — **moodley brand identity**
Client — **Amisola Immobilien AG**
Creative Director — **Gerd Schicketanz**
Art Director — **Gerd Schicketanz**
Graphic Designer — **Katharina Hölzl**
Text & Book Concepts — **Michael Endlicher**
Photographer — **Andreas Balon, gregortitze.com**

DESCRIPTION

For centuries, the Fleischmarkt has been
a traditional center of attraction for European
trade. At the start of the 20th century,
the "Residenzpalast" was built — an architectural
milestone with visionary constructional
techniques. Now, it has been converted to an
office building with large-scale historic façades.
moodley brand identity developed a suitable
corporate design and communication
and marketing concept for the available
office spaces and future tenants.

The inspiration: the historic architecture
and futuristic, high-end equipment of the
office spaces. The focus lies on large-scale
photographs, the reserved colour scape and the
use of a contrasting type face create an arc of
suspense between living architectural history
and contemporary, representative interior
design. Short: Fleischmarkt 1 — premium
offices. Inspiring history. Spurring future.

FLEISCH MARKT 1
Premium Offices

Top	max. Raumhöhe	Nutzfläche (m²)	Hauptzugang
Top 501	3,20	358,23	Fleischmarkt 1 (Stiege 1)
Top 502	3,20	717,01	Fleischmarkt 1 (Stiege 1)
Top 503	3,20	655,63	Fleischmarkt 3 (Stiege 4)
Summe			
5. Obergeschoss		1.730,87	

5.Obergeschoss
ohne Zwischenwände

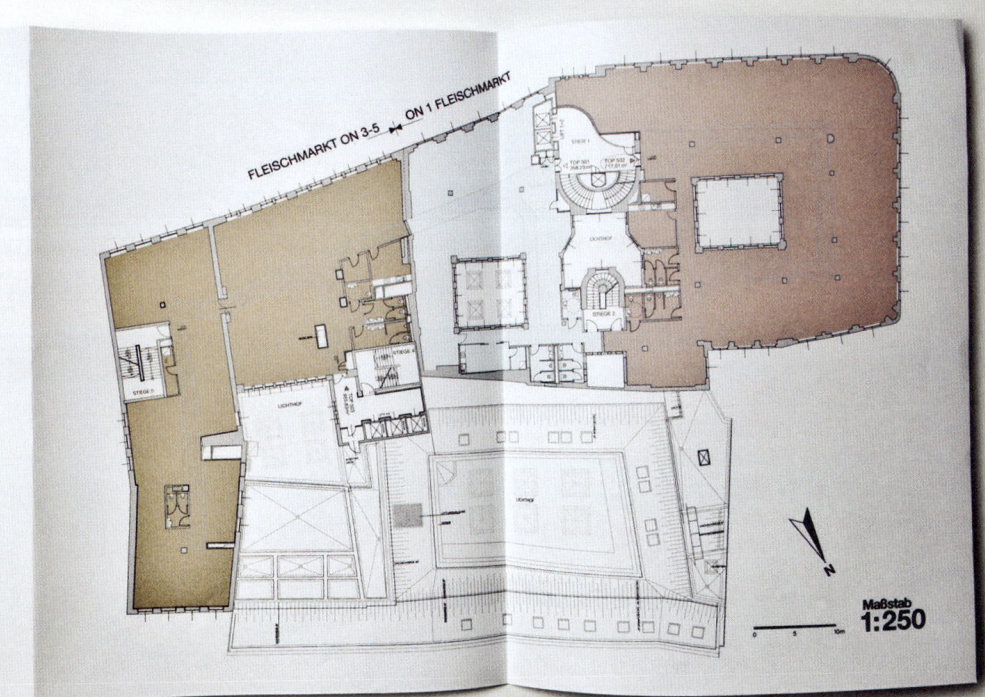

ORCHARD PIPER: WASHINGTON STREET

CREDITS

Studio — **FABIO ONGARATO DESIGN**
Creative Director — **Fabio Ongarato**
Designer — **Paul Tisdell**
Photographer — **Adrian Gant**

DESCRIPTION

Positioned as "One of a Kind," Orchard Piper's Washington Street Development presents a new benchmark in luxury property. The sales and marketing campaign is a celebration of "The Makers" — a team of international artisans and designers whose contribution to Washington Street is an authentic, architectural statement. The campaign, communicating bespoke, uncompromising quality and rarity in the market was applied on-line, across printed collateral, project gallery and a highly targeted media strategy.

ORCHARD PIPER
WASHINGTON STREET

THE GARDEN RESIDENCES
APARTMENT I

ORCHARD PIPER
WASHINGTON STREET

APPENDIX TWO
BESPOKE LIVING

ORCHARD PIPER
WASHINGTON STREET

APPENDIX ONE
LUXURY AS STANDARD

ORCHARD PIPER
WASHINGTON STREET

ORCHARD PIPER

ORCHARD PIPER

ORCHARD PIPER

ORCHARD PIPER

ORCHARD PIPER

ORCHARD PIPER

ORCH

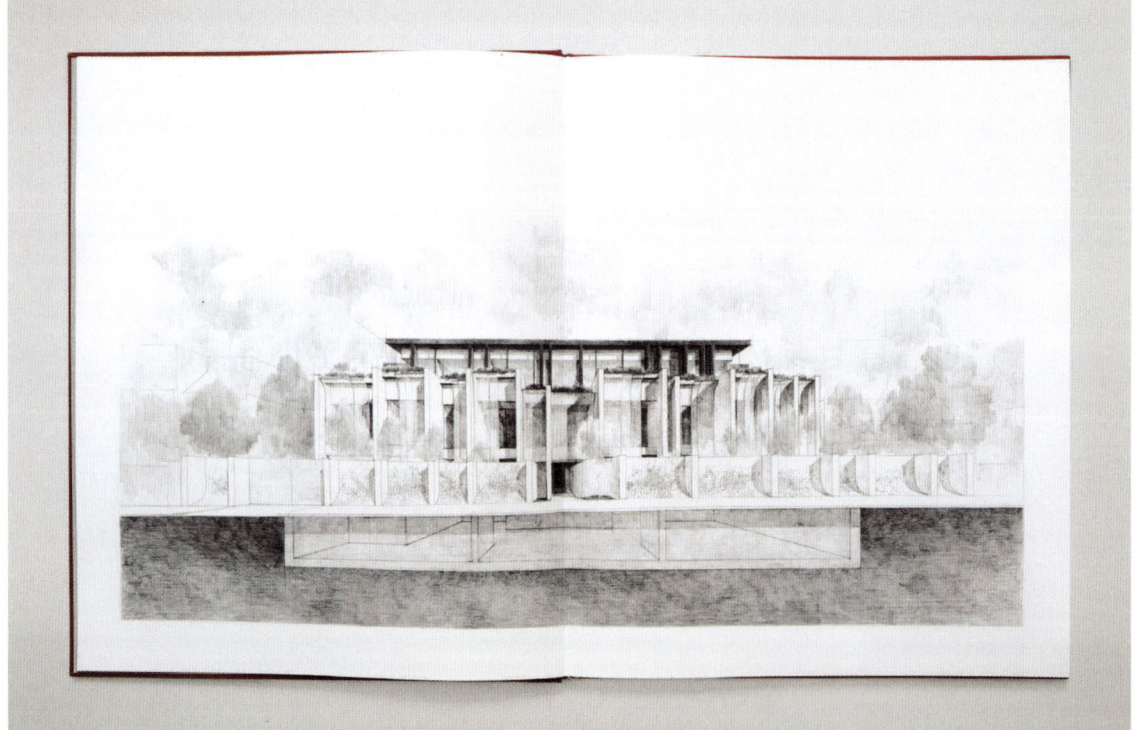

INDEX

PROJECT INDEX

SUBMITTER INDEX

Published in 2016.

Curated, Edited and Designed by
RHED Publishers

We would like to thank all the designers and companies
involved in the compilation of this book. This project
would not have been accomplished without their
significant contribution. We would also like to
express our gratitude to all the producers for their
invaluable opinions and assistance all this time. This
book's successful completion also owes a great deal to
many professionals in the creative industry who have
provided precious insights and comments. Lastly, to
many other whose names, though not credited, who
have made a big impact on our work, we thank you for
your continuous support the whole time.